TOURO COLLEGE LIBRARY
Bay Shore Campus

MONOGRAPHS ON STATISTICS AND APPLIED PROBABILITY

General Editors

D.R. Cox, V. Isham, N. Keiding, T. Louis, N. Reid, R. Tibshirani, and H. Tong

1 Stochastic Population Models in Ecology and Epidemiology *M.S. Barlett* (1960)
2 Queues *D.R. Cox and W.L. Smith* (1961)
3 Monte Carlo Methods *J.M. Hammersley and D.C. Handscomb* (1964)
4 The Statistical Analysis of Series of Events *D.R. Cox and P.A.W. Lewis* (1966)
5 Population Genetics *W.J. Ewens* (1969)
6 Probability, Statistics and Time *M.S. Barlett* (1975)
7 Statistical Inference *S.D. Silvey* (1975)
8 The Analysis of Contingency Tables *B.S. Everitt* (1977)
9 Multivariate Analysis in Behavioural Research *A.E. Maxwell* (1977)
10 Stochastic Abundance Models *S. Engen* (1978)
11 Some Basic Theory for Statistical Inference *E.J.G. Pitman* (1979)
12 Point Processes *D.R. Cox and V. Isham* (1980)
13 Identification of Outliers *D.M. Hawkins* (1980)
14 Optimal Design *S.D. Silvey* (1980)
15 Finite Mixture Distributions *B.S. Everitt and D.J. Hand* (1981)
16 Classification *A.D. Gordon* (1981)
17 Distribution-Free Statistical Methods, 2nd edition *J.S. Maritz* (1995)
18 Residuals and Influence in Regression *R.D. Cook and S. Weisberg* (1982)
19 Applications of Queueing Theory, 2nd edition *G.F. Newell* (1982)
20 Risk Theory, 3rd edition *R.E. Beard, T. Pentikäinen and E. Pesonen* (1984)
21 Analysis of Survival Data *D.R. Cox and D. Oakes* (1984)
22 An Introduction to Latent Variable Models *B.S. Everitt* (1984)
23 Bandit Problems *D.A. Berry and B. Fristedt* (1985)
24 Stochastic Modelling and Control *M.H.A. Davis and R. Vinter* (1985)
25 The Statistical Analysis of Composition Data *J. Aitchison* (1986)
26 Density Estimation — for Statistics and Data Analysis *B.W. Silverman* (1986)
27 Regression Analysis with Applications *G.B. Wetherill* (1986)
28 Sequential Methods in Statistics, 3rd edition
G.B. Wetherill and K.D. Glazebrook (1986)
29 Tensor Methods in Statistics *P. McCullagh* (1987)
30 Transformation and Weighting in Regression
R.J. Carroll and D. Ruppert (1988)
31 Asymptotic Techniques for Use in Statistics
O.E. Bandorff-Nielsen and D.R. Cox (1989)
32 Analysis of Binary Data, 2nd edition *D.R. Cox and E.J. Snell* (1989)

33 Analysis of Infectious Disease Data *N.G. Becker* (1989)
34 Design and Analysis of Cross-Over Trials *B. Jones and M.G. Kenward* (1989)
35 Empirical Bayes Methods, 2nd edition *J.S. Maritz and T. Lwin* (1989)
36 Symmetric Multivariate and Related Distributions
K.T. Fang, S. Kotz and K.W. Ng (1990)
37 Generalized Linear Models, 2nd edition *P. McCullagh and J.A. Nelder* (1989)
38 Cyclic and Computer Generated Designs, 2nd edition
J.A. John and E.R. Williams (1995)
39 Analog Estimation Methods in Econometrics *C.F. Manski* (1988)
40 Subset Selection in Regression *A.J. Miller* (1990)
41 Analysis of Repeated Measures *M.J. Crowder and D.J. Hand* (1990)
42 Statistical Reasoning with Imprecise Probabilities *P. Walley* (1991)
43 Generalized Additive Models *T.J. Hastie and R.J. Tibshirani* (1990)
44 Inspection Errors for Attributes in Quality Control
N.L. Johnson, S. Kotz and X, Wu (1991)
45 The Analysis of Contingency Tables, 2nd edition *B.S. Everitt* (1992)
46 The Analysis of Quantal Response Data *B.J.T. Morgan* (1992)
47 Longitudinal Data with Serial Correlation—A state-space approach
R.H. Jones (1993)
48 Differential Geometry and Statistics *M.K. Murray and J.W. Rice* (1993)
49 Markov Models and Optimization *M.H.A. Davis* (1993)
50 Networks and Chaos—statistical and probabilistic aspects
O.E. Barndorff-Nielsen, J.L. Jensen and W.S. Kendall (1993)
51 Number-Theoretic Methods in Statistics *K.-T. Fang and Y. Wang* (1994)
52 Inference and Asymptotics *O.E. Barndorff-Nielsen and D.R. Cox* (1994)
53 Practical Risk Theory for Actuaries
C.D. Daykin, T. Pentikäinen and M. Pesonen (1994)
54 Biplots *J.C. Gower and D.J. Hand* (1996)
55 Predictive Inference—An introduction *S. Geisser* (1993)
56 Model-Free Curve Estimation *M.E. Tarter and M.D. Lock* (1993)
57 An Introduction to the Bootstrap *B. Efron and R.J. Tibshirani* (1993)
58 Nonparametric Regression and Generalized Linear Models
P.J. Green and B.W. Silverman (1994)
59 Multidimensional Scaling *T.F. Cox and M.A.A. Cox* (1994)
60 Kernel Smoothing *M.P. Wand and M.C. Jones* (1995)
61 Statistics for Long Memory Processes *J. Beran* (1995)
62 Nonlinear Models for Repeated Measurement Data
M. Davidian and D.M. Giltinan (1995)
63 Measurement Error in Nonlinear Models
R.J. Carroll, D. Rupert and L.A. Stefanski (1995)
64 Analyzing and Modeling Rank Data *J.J. Marden* (1995)
65 Time Series Models—In econometrics, finance and other fields
D.R. Cox, D.V. Hinkley and O.E. Barndorff-Nielsen (1996)

66 Local Polynomial Modeling and its Applications *J. Fan and I. Gijbels* (1996)
67 Multivariate Dependencies—Models, analysis and interpretation *D.R. Cox and N. Wermuth* (1996)
68 Statistical Inference—Based on the likelihood *A. Azzalini* (1996)
69 Bayes and Empirical Bayes Methods for Data Analysis *B.P. Carlin and T.A Louis* (1996)
70 Hidden Markov and Other Models for Discrete-Valued Time Series *I.L. Macdonald and W. Zucchini* (1997)
71 Statistical Evidence—A likelihood paradigm *R. Royall* (1997)
72 Analysis of Incomplete Multivariate Data *J.L. Schafer* (1997)
73 Multivariate Models and Dependence Concepts *H. Joe* (1997)
74 Theory of Sample Surveys *M.E. Thompson* (1997)
75 Retrial Queues *G. Falin and J.G.C. Templeton* (1997)
76 Theory of Dispersion Models *B. Jørgensen* (1997)
77 Mixed Poisson Processes *J. Grandell* (1997)
78 Variance Components Estimation—Mixed models, methodologies and applications *P.S.R.S. Rao* (1997)
79 Bayesian Methods for Finite Population Sampling *G. Meeden and M. Ghosh* (1997)
80 Stochastic Geometry—Likelihood and computation *O.E. Barndorff-Nielsen, W.S. Kendall and M.N.M. van Lieshout* (1998)
81 Computer-Assisted Analysis of Mixtures and Applications— Meta-analysis, disease mapping and others *D. Böhning* (1999)
82 Classification, 2nd edition *A.D. Gordon* (1999)
83 Semimartingales and their Statistical Inference *B.L.S. Prakasa Rao* (1999)
84 Statistical Aspects of BSE and vCJD—Models for Epidemics *C.A. Donnelly and N.M. Ferguson* (1999)
85 Set-Indexed Martingales *G. Ivanoff and E. Merzbach* (2000)
86 The Theory of the Design of Experiments *D.R. Cox and N. Reid* (2000)

The Theory of the Design of Experiments

D.R. COX
Honorary Fellow
Nuffield College
Oxford, UK

AND

N. REID
Professor of Statistics
University of Toronto, Canada

CHAPMAN & HALL/CRC
Boca Raton   London   New York   Washington, D.C.

Library of Congress Cataloging-in-Publication Data

Cox, D. R. (David Roxbee)
　　The theory of the design of experiments / D. R. Cox, N. Reid.
　　　　p.　cm. — (Monographs on statistics and applied probability ; 86)
　　Includes bibliographical references and index.
　　ISBN 1-58488-195-X (alk. paper)
　　1. Experimental design. I. Reid, N. II.Title. III. Series.
　QA279 .C73 2000
　001.4′34—dc21　　　　　　　　　　　　　　　　　　　　　　00-029529
　　　　　　　　　　　　　　　　　　　　　　　　　　　　　　　　　CIP

This book contains information obtained from authentic and highly regarded sources. Reprinted material is quoted with permission, and sources are indicated. A wide variety of references are listed. Reasonable efforts have been made to publish reliable data and information, but the author and the publisher cannot assume responsibility for the validity of all materials or for the consequences of their use.

Neither this book nor any part may be reproduced or transmitted in any form or by any means, electronic or mechanical, including photocopying, microfilming, and recording, or by any information storage or retrieval system, without prior permission in writing from the publisher.

The consent of CRC Press LLC does not extend to copying for general distribution, for promotion, for creating new works, or for resale. Specific permission must be obtained in writing from CRC Press LLC for such copying.

Direct all inquiries to CRC Press LLC, 2000 N.W. Corporate Blvd., Boca Raton, Florida 33431.

Trademark Notice: Product or corporate names may be trademarks or registered trademarks, and are used only for identification and explanation, without intent to infringe.

Visit the CRC Press Web site at www.crcpress.com

© 2000 by Chapman & Hall/CRC

No claim to original U.S. Government works
International Standard Book Number 1-58488-195-X
Library of Congress Card Number 00-029529
Printed in the United States of America　　2 3 4 5 6 7 8 9 0
Printed on acid-free paper

Contents

Preface

1 Some general concepts   1
   1.1 Types of investigation   1
   1.2 Observational studies   3
   1.3 Some key terms   4
   1.4 Requirements in design   5
   1.5 Interplay between design and analysis   6
   1.6 Key steps in design   7
   1.7 A simplified model   11
   1.8 A broader view   11
   1.9 Bibliographic notes   14
   1.10 Further results and exercises   15

2 Avoidance of bias   19
   2.1 General remarks   19
   2.2 Randomization   19
   2.3 Retrospective adjustment for bias   29
   2.4 Some more on randomization   32
   2.5 More on causality   34
   2.6 Bibliographic notes   36
   2.7 Further results and exercises   37

3 Control of haphazard variation   41
   3.1 General remarks   41
   3.2 Precision improvement by blocking   42
   3.3 Matched pairs   43
   3.4 Randomized block design   48
   3.5 Partitioning sums of squares   53
   3.6 Retrospective adjustment for improving precision   57
   3.7 Special models of error variation   61

	3.8	Bibliographic notes	62
	3.9	Further results and exercises	62
4	**Specialized blocking techniques**		**65**
	4.1	Latin squares	65
	4.2	Incomplete block designs	70
	4.3	Cross-over designs	85
	4.4	Bibliographic notes	95
	4.5	Further results and exercises	95
5	**Factorial designs: basic ideas**		**99**
	5.1	General remarks	99
	5.2	Example	101
	5.3	Main effects and interactions	102
	5.4	Example: continued	109
	5.5	Two level factorial systems	110
	5.6	Fractional factorials	116
	5.7	Example	120
	5.8	Bibliographic notes	122
	5.9	Further results and exercises	123
6	**Factorial designs: further topics**		**127**
	6.1	General remarks	127
	6.2	Confounding in 2^k designs	127
	6.3	Other factorial systems	131
	6.4	Split plot designs	140
	6.5	Nonspecific factors	144
	6.6	Designs for quantitative factors	149
	6.7	Taguchi methods	157
	6.8	Conclusion	160
	6.9	Bibliographic notes	162
	6.10	Further results and exercises	163
7	**Optimal design**		**169**
	7.1	General remarks	169
	7.2	Some simple examples	169
	7.3	Some general theory	173
	7.4	Other optimality criteria	176
	7.5	Algorithms for design construction	177
	7.6	Nonlinear design	178

7.7	Space-filling designs	181
7.8	Bayesian design	182
7.9	Optimality of traditional designs	186
7.10	Bibliographic notes	186
7.11	Further results and exercises	188

8 Some additional topics — 193

8.1	Scale of effort	193
8.2	Adaptive designs	201
8.3	Sequential regression design	208
8.4	Designs for one-dimensional error structure	209
8.5	Spatial designs	215
8.6	Bibliographic notes	219
8.7	Further results and exercises	221

A Statistical analysis — 225

A.1	Introduction	225
A.2	Linear model	226
A.3	Analysis of variance	238
A.4	More general models; maximum likelihood	244
A.5	Bibliographic notes	244
A.6	Further results and exercises	245

B Some algebra — 249

B.1	Introduction	249
B.2	Group theory	249
B.3	Galois fields	254
B.4	Finite geometries	258
B.5	Difference sets	260
B.6	Hadamard matrices	261
B.7	Orthogonal arrays	262
B.8	Coding theory	263
B.9	Bibliographic notes	264
B.10	Further results and exercises	265

C Computational issues — 267

C.1	Introduction	267
C.2	Overview	268
C.3	Randomized block experiment from Chapter 3	273
C.4	Analysis of block designs in Chapter 4	280

C.5	Examples from Chapter 5	285
C.6	Examples from Chapter 6	292
C.7	Bibliographic notes	297

References   299

List of tables   311

Author index   313

Index   317

Preface

This book is an account of the major topics in the design of experiments, with particular emphasis on the key concepts involved and on the statistical structure associated with these concepts. While design of experiments is in many ways a very well developed area of statistics, it often receives less emphasis than methods of analysis in a programme of study in statistics.

We have written for a general audience concerned with statistics in experimental fields and with some knowledge of and interest in theoretical issues. The mathematical level is mostly elementary; occasional passages using more advanced ideas can be skipped or omitted without inhibiting understanding of later passages. Some specialized parts of the subject have extensive and specialized literatures, a few examples being incomplete block designs, mixture designs, designs for large variety trials, designs based on spatial stochastic models and designs constructed from explicit optimality requirements. We have aimed to give relatively brief introductions to these subjects eschewing technical detail.

To motivate the discussion we give outline *Illustrations* taken from a range of areas of application. In addition we give a limited number of *Examples*, mostly taken from the literature, used for the different purpose of showing detailed methods of analysis without much emphasis on specific subject-matter interpretation.

We have written a book about design not about analysis, although, as has often been pointed out, the two phases are inexorably interrelated. Therefore it is, in particular, not a book on the linear statistical model or that related but distinct art form the analysis of variance table. Nevertheless these topics enter and there is a dilemma in presentation. What do we assume the reader knows about these matters? We have solved this problem uneasily by somewhat downplaying analysis in the text, by assuming whatever is necessary for the section in question and by giving a review as an Appendix. Anyone using the book as a basis for a course of lectures will need to consider carefully what the prospective students are likely to understand about the linear model and to supplement the text appropriately. While the arrangement of chapters represents a logical progression of ideas, if interest is focused on a particular field of application it will be reasonable to omit certain parts or to take the material in a different order.

If defence of a book on the theory of the subject is needed it is this. Successful application of these ideas hinges on adapting general principles to the special constraints of individual applications. Thus experience suggests that while it is useful to know about special designs, balanced incomplete block designs for instance, it is rather rare that they can be used directly. More commonly they need some adaptation to accommodate special circumstances and to do this effectively demands a solid theoretical base.

This book has been developed from lectures given at Cambridge, Birkbeck College London, Vancouver, Toronto and Oxford. We are grateful to Amy Berrington, Mario Cortina Borja, Christl Donnelly, Peter Kupchak, Rahul Mukerjee, John Nelder, Rob Tibshirani and especially Grace Yun Yi for helpful comments on a preliminary version.

D.R. Cox and N. Reid
Oxford and Toronto
January 2000

CHAPTER 1

Some general concepts

1.1 Types of investigation

This book is about the design of experiments. The word *experiment* is used in a quite precise sense to mean an investigation where the system under study is under the control of the investigator. This means that the individuals or material investigated, the nature of the treatments or manipulations under study and the measurement procedures used are all settled, in their important features at least, by the investigator.

By contrast in an observational study some of these features, and in particular the allocation of individuals to treatment groups, are outside the investigator's control.

Illustration. In a randomized clinical trial patients meeting clearly defined eligibility criteria and giving informed consent are assigned by the investigator by an impersonal method to one of two or more treatment arms, in the simplest case either to a new treatment, T, or to a standard treatment or control, C, which might be the best current therapy or possibly a placebo treatment. The patients are followed for a specified period and one or more measures of response recorded.

In a comparable observational study, data on the response variables might be recorded on two groups of patients, some of whom had received T and some C; the data might, for example, be extracted from a database set up during the normal running of a hospital clinic. In such a study, however, it would be unknown why each particular patient had received the treatment he or she had.

The form of data might be similar or even almost identical in the two contexts; the distinction lies in the firmness of the interpretation that can be given to the apparent differences in response between the two groups of patients.

Illustration. In an agricultural field trial, an experimental field is divided into plots of a size and shape determined by the investigator, subject to technological constraints. To each plot is assigned one of a number of fertiliser treatments, often combinations of various amounts of the basic constituents, and yield of product is measured.

In a comparable observational study of fertiliser practice a survey of farms or fields would give data on quantities of fertiliser used and yield, but the issue of why each particular fertiliser combination had been used in each case would be unclear and certainly not under the investigator's control.

A common feature of these and other similar studies is that the objective is a comparison, of two medical treatments in the first example, and of various fertiliser combinations in the second. Many investigations in science and technology have this form. In very broad terms, in technological experiments the treatments under comparison have a very direct interest, whereas in scientific experiments the treatments serve to elucidate the nature of some phenomenon of interest or to test some research hypothesis. We do not, however, wish to emphasize distinctions between science and technology.

We translate the objective into that of comparing the responses among the different treatments. An experiment and an observational study may have identical objectives; the distinction between them lies in the confidence to be put in the interpretation.

Investigations done wholly or primarily in the laboratory are usually experimental. Studies of social science issues in the context in which they occur in the real world are usually inevitably observational, although sometimes elements of an experimental approach may be possible. Industrial studies at pilot plant level will typically be experimental whereas at a production level, while experimental approaches are of proved fruitfulness especially in the process industries, practical constraints may force some deviation from what is ideal for clarity of interpretation.

Illustration. In a survey of social attitudes a panel of individuals might be interviewed say every year. This would be an observational study designed to study and if possible explain changes of attitude over time. In such studies panel attrition, i.e. loss of respondents for one reason or another, is a major concern. One way of reducing attrition may be to offer small monetary payments a few days before the interview is due. An experiment on the effectiveness of this could take the form of randomizing individuals between one of two treatments, a monetary payment or no monetary payment. The response would be the successful completion or not of an interview.

1.2 Observational studies

While in principle the distinction between experiments and observational studies is clear cut and we wish strongly to emphasize its importance, nevertheless in practice the distinction can sometimes become blurred. Therefore we comment briefly on various forms of observational study and on their closeness to experiments.

It is helpful to distinguish between a prospective longitudinal study (cohort study), a retrospective longitudinal study, a cross-sectional study, and the secondary analysis of data collected for some other, for example, administrative purpose.

In a *prospective study* observations are made on individuals at entry into the study, the individuals are followed forward in time, and possible response variables recorded for each individual. In a *retrospective study* the response is recorded at entry and an attempt is made to look backwards in time for possible explanatory features. In a *cross-sectional study* each individual is observed at just one time point. In all these studies the investigator may have substantial control not only over which individuals are included but also over the measuring processes used. In a *secondary analysis* the investigator has control only over the inclusion or exclusion of the individuals for analysis.

In a general way the four possibilities are in decreasing order of effectiveness, the prospective study being closest to an experiment; they are also in decreasing order of cost.

Thus retrospective studies are subject to biases of recall but may often yield results much more quickly than corresponding prospective studies. In principle at least, observations taken at just one time point are likely to be less enlightening than those taken over time. Finally secondary analysis, especially of some of the large databases now becoming so common, may appear attractive. The quality of such data may, however, be low and there may be major difficulties in disentangling effects of different explanatory features, so that often such analyses are best regarded primarily as ways of generating ideas for more detailed study later.

In epidemiological applications, a retrospective study is often designed as a *case-control* study, whereby groups of patients with a disease or condition (cases), are compared to a hopefully similar group of disease-free patients on their exposure to one or more risk factors.

1.3 Some key terms

We shall return later to a more detailed description of the types of experiment to be considered but for the moment it is enough to consider three important elements to an experiment, namely the experimental units, the treatments and the response. A schematic version of an experiment is that there are a number of different treatments under study, the investigator assigns one treatment to each experimental unit and observes the resulting response.

Experimental units are essentially the patients, plots, animals, raw material, etc. of the investigation. More formally they correspond to the smallest subdivision of the experimental material such that any two different experimental units might receive different treatments.

Illustration. In some experiments in opthalmology it might be sensible to apply different treatments to the left and to the right eye of each patient. Then an experimental unit would be an eye, that is each patient would contribute two experimental units.

The treatments are clearly defined procedures one of which is to be applied to each experimental unit. In some cases the treatments are an unstructured set of two or more qualitatively different procedures. In others, including many investigations in the physical sciences, the treatments are defined by the levels of one or more quantitative variables, such as the amounts per square metre of the constituents nitrogen, potash and potassium, in the illustration in Section 1.1.

The response measurement specifies the criterion in terms of which the comparison of treatments is to be effected. In many applications there will be several such measures.

This simple formulation can be amplified in various ways. The same physical material can be used as an experimental unit more than once. If the treatment structure is complicated the experimental unit may be different for different components of treatment. The response measured may be supplemented by measurements on other properties, called *baseline variables*, made before allocation to treatment, and on *intermediate variables* between the baseline variables and the ultimate response.

Illustrations. In clinical trials there will typically be available numerous baseline variables such as age at entry, gender, and specific properties relevant to the disease, such as blood pressure, etc., all to be measured before assignment to treatment. If the key response is time to death, or more generally time to some critical event in the progression of the disease, intermediate variables might be properties measured during the study which monitor or explain the progression to the final response.

In an agricultural field trial possible baseline variables are chemical analyses of the soil in each plot and the yield on the plot in the previous growing season, although, so far as we are aware, the effectiveness of such variables as an aid to experimentation is limited. Possible intermediate variables are plant density, the number of plants per square metre, and assessments of growth at various intermediate points in the growing season. These would be included to attempt explanation of the reasons for the effect of fertiliser on yield of final product.

1.4 Requirements in design

The objective in the type of experiment studied here is the comparison of the effect of treatments on response. This will typically be assessed by estimates and confidence limits for the magnitude of treatment differences. Requirements on such estimates are essentially as follows. First systematic errors, or biases, are to be avoided. Next the effect of random errors should so far as feasible be minimized. Further it should be possible to make reasonable assessment of the magnitude of random errors, typically via confidence limits for the comparisons of interest. The scale of the investigation should be such as to achieve a useful but not unnecessarily high level of precision. Finally advantage should be taken of any special structure in the treatments, for example when these are specified by combinations of factors.

The relative importance of these aspects is different in different fields of study. For example in large clinical trials to assess relatively small differences in treatment efficacy, avoidance of systematic error is a primary issue. In agricultural field trials, and probably more generally in studies that do not involve human subjects, avoidance of bias, while still important, is not usually the aspect of main concern.

These objectives have to be secured subject to the practical constraints of the situation under study. The designs and considerations developed in this book have often to be adapted or modified to meet such constraints.

1.5 Interplay between design and analysis

There is a close connection between design and analysis in that an objective of design is to make both analysis and interpretation as simple and clear as possible. Equally, while some defects in design may be corrected by more elaborate analysis, there is nearly always some loss of security in the interpretation, i.e. in the underlying subject-matter meaning of the outcomes.

The choice of detailed model for analysis and interpretation will often involve subject-matter considerations that cannot readily be discussed in a general book such as this. Partly but not entirely for this reason we concentrate here on the analysis of continuously distributed responses via models that are usually linear, leading to analyses quite closely connected with the least-squares analysis of the normal theory linear model. One intention is to show that such default analyses follow from a single set of assumptions common to the majority of the designs we shall consider. In this rather special sense, the model for analysis is determined by the design employed. Of course we do not preclude the incorporation of special subject-matter knowledge and models where appropriate and indeed this may be essential for interpretation.

There is a wider issue involved especially when a number of different response variables are measured and underlying interpretation is the objective rather than the direct estimation of treatment differences. It is sensible to try to imagine the main patterns of response that are likely to arise and to consider whether the information will have been collected to allow the interpretation of these. This is a broader issue than that of reviewing the main scheme of analysis to be used. Such consideration must always be desirable; it is, however, considerably less than a prior commitment to a very detailed approach to analysis.

Two terms quite widely used in discussions of the design of experiments are *balance* and *orthogonality*. Their definition depends a bit on context but broadly balance refers to some strong symmetry in the combinatorial structure of the design, whereas orthogonality refers to special simplifications of analysis and achievement of efficiency consequent on such balance.

For example, in Chapter 3 we deal with designs for a number of treatments in which the experimental units are arranged in blocks. The design is balanced if each treatment occurs in each block the same number of times, typically once. If a treatment occurs once in some blocks and twice or not at all in others the design is considered unbalanced. On the other hand, in the context of balanced incomplete block designs studied in Section 4.2 the word balance refers to an extended form of symmetry.

In analyses involving a linear model, and most of our discussion centres on these, two types of effect are orthogonal if the relevant columns of the matrix defining the linear model are orthogonal in the usual algebraic sense. One consequence is that the least squares estimates of one of the effects are unchanged if the other type of effect is omitted from the model. For orthogonality some kinds of balance are sufficient but not necessary. In general statistical theory there is an extended notion of orthogonality based on the Fisher information matrix and this is relevant when maximum likelihood analysis of more complicated models is considered.

1.6 Key steps in design

1.6.1 General remarks

Clearly the single most important aspect of design is a purely substantive, i.e. subject-matter, one. The issues addressed should be interesting and fruitful. Usually this means examining one or more well formulated questions or research hypotheses, for example a speculation about the process underlying some phenomenon, or the clarification and explanation of earlier findings. Some investigations may have a less focused objective. For example, the initial phases of a study of an industrial process under production conditions may have the objective of identifying which few of a large number of potential influences are most important. The methods of Section 5.6 are aimed at such situations, although they are probably atypical and in most cases the more specific the research question the better.

In principle therefore the general objectives lead to the following more specific issues. First the experimental units must be defined and chosen. Then the treatments must be clearly defined. The variables to be measured on each unit must be specified and finally the size of the experiment, in particular the number of experimental units, has to be decided.

1.6.2 Experimental units

Issues concerning experimental units are to some extent very specific to each field of application. Some points that arise fairly generally and which influence the discussion in this book include the following.

Sometimes, especially in experiments with a technological focus, it is useful to consider the population of ultimate interest and the population of accessible individuals and to aim at conclusions that will bridge the inevitable gap between these. This is linked to the question of whether units should be chosen to be as uniform as possible or to span a range of circumstances. Where the latter is sensible it will be important to impose a clear structure on the experimental units; this is connected with the issue of the choice of baseline measurements.

Illustration. In agricultural experimentation with an immediate objective of making recommendations to farmers it will be important to experiment in a range of soil and weather conditions; a very precise conclusion in one set of conditions may be of limited value. Interpretation will be much simplified if the same basic design is used at each site. There are somewhat similar considerations in some clinical trials, pointing to the desirability of multi-centre trials even if a trial in one centre would in principle be possible.

By contrast in experiments aimed at elucidating the nature of certain processes or mechanisms it will usually be best to choose units likely to show the effect in question in as striking a form as possible and to aim for a high degree of uniformity across units.

In some contexts the same individual animal, person or material may be used several times as an experimental unit; for example in a psychological experiment it would be common to expose the same subject to various conditions (treatments) in one session.

It is important in much of the following discussion and in applications to distinguish between experimental units and observations. The key notion is that different experimental units must in principle be capable of receiving different treatments.

KEY STEPS IN DESIGN 9

Illustration. In an industrial experiment on a batch process each separate batch of material might form an experimental unit to be processed in a uniform way, separate batches being processed possibly differently. On the product of each batch many samples may be taken to measure, say purity of the product. The number of observations of purity would then be many times the number of experimental units. Variation between repeat observations within a batch measures sampling variability and internal variability of the process. Precision of the comparison of treatments is, however, largely determined by, and must be estimated from, variation between batches receiving the same treatment. In our theoretical treatment that follows the number of batches is thus the relevant total "sample" size.

1.6.3 Treatments

The simplest form of experiment compares a new treatment or manipulation, T, with a control, C. Even here care is needed in applications. In principle T has to be specified with considerable precision, including many details of its mode of implementation. The choice of control, C, may also be critical. In some contexts several different control treatments may be desirable. Ideally the control should be such as to isolate precisely that aspect of T which it is the objective to examine.

Illustration. In a clinical trial to assess a new drug, the choice of control may depend heavily on the context. Possible choices of control might be no treatment, a placebo treatment, i.e. a substance superficially indistinguishable from T but known to be pharmacologically inactive, or the best currently available therapy. The choice between placebo and best available treatment may in some clinical trials involve difficult ethical decisions.

In more complex situations there may be a collection of qualitatively different treatments $T_1, \ldots, T_v$. More commonly the treatments may have factorial structure, i.e. be formed from combinations of levels of subtreatments, called *factors*. We defer detailed study of the different kinds of factor and the design of factorial experiments until Chapter 5, noting that sensible use of the principle of examining several factors together in one study is one of the most powerful ideas in this subject.

1.6.4 Measurements

The choice of appropriate variables for measurement is a key aspect of design in the broad sense. The nature of measurement processes and their associated potentiality for error, and the different kinds of variable that can be measured and their purposes are central issues. Nevertheless these issues fall outside the scope of the present book and we merely note three broad types of variable, namely baseline variables describing the experimental units before application of treatments, intermediate variables and response variables, in a medical context often called end-points.

Intermediate variables may serve different roles. Usually the more important is to provide some provisional explanation of the process that leads from treatment to response. Other roles are to check on the absence of untoward interventions and, sometimes, to serve as surrogate response variables when the primary response takes a long time to obtain.

Sometimes the response on an experimental unit is in effect a time trace, for example of the concentrations of one or more substances as transient functions of time after some intervention. For our purposes we suppose such responses replaced by one or more summary measures, such as the peak response or the area under the response-time curve.

Clear decisions about the variables to be measured, especially the response variables, are crucial.

1.6.5 Size of experiment

Some consideration virtually always has to be given to the number of experimental units to be used and, where subsampling of units is employed, to the number of repeat observations per unit. A balance has to be struck between the marginal cost per experimental unit and the increase in precision achieved per additional unit. Except in rare instances where these costs can both be quantified, a decision on the size of experiment is bound be largely a matter of judgement and some of the more formal approaches to determining the size of the experiment have spurious precision. It is, however, very desirable to make an advance approximate calculation of the precision likely to be achieved. This gives some protection against wasting resources on unnecessary precision or, more commonly, against undertaking investigations which will be of such low precision that useful conclusions are very unlikely. The same calculations are advisable when, as is quite common in some fields, the maximum size of the experiment is set by constraints outside the control of the investigator. The issue is then most commonly to decide whether the resources are sufficient to yield enough precision to justify proceeding at all.

1.7 A simplified model

The formulation of experimental design that will largely be used in this book is as follows. There are given n experimental units, $U_1, \ldots, U_n$ and v treatments, $T_1, \ldots, T_v$; one treatment is applied to each unit as specified by the investigator, and one response Y measured on each unit. The objective is to specify procedures for allocating treatments to units and for the estimation of the differences between treatments in their effect on response.

This is a very limited version of the broader view of design sketched above. The justification for it is that many of the valuable specific designs are accommodated in this framework, whereas the wider considerations sketched above are often so subject-specific that it is difficult to give a general theoretical discussion.

It is, however, very important to recall throughout that the path between the choice of a unit and the measurement of final response may be a long one in time and in other respects and that random and systematic error may arise at many points. Controlling for random error and aiming to eliminate systematic error is thus not a single step matter as might appear in our idealized model.

1.8 A broader view

The discussion above and in the remainder of the book concentrates on the integrity of individual experiments. Yet investigations are rarely if ever conducted in isolation; one investigation almost inevitably suggests further issues for study and there is commonly the need to establish links with work related to the current problems, even if only rather distantly. These are important matters but again are difficult to incorporate into formal theoretical discussion.

If a given collection of investigations estimate formally the same contrasts, the statistical techniques for examining mutual consistency of the different estimates and, subject to such consistency, of combining the information are straightforward. Difficulties come more from the choice of investigations for inclusion, issues of genuine comparability and of the resolution of apparent inconsistencies.

While we take the usual objective of the investigation to be the comparison of responses from different treatments, sometimes there is a more specific objective which has an impact on the design to be employed.

Illustrations. In some kinds of investigation in the chemical process industries, the treatments correspond to differing concentrations of various reactants and to variables such as pressure, temperature, etc. For some purposes it may be fruitful to regard the objective as the determination of conditions that will optimize some criterion such as yield of product or yield of product per unit cost. Such an explicitly formulated purpose, if adopted as the sole objective, will change the approach to design.

In selection programmes for, say, varieties of wheat, the investigation may start with a very large number of varieties, possibly several hundred, for comparison. A certain fraction of these are chosen for further study and in a third phase a small number of varieties are subject to intensive study. The initial stage has inevitably very low precision for individual comparisons and analysis of the design strategy to be followed best concentrates on such issues as the proportion of varieties to be chosen at each phase, the relative effort to be devoted to each phase and in general on the properties of the whole process and the properties of the varieties ultimately selected rather than on the estimation of individual differences.

In the pharmaceutical industry clinical trials are commonly defined as Phase I, II or III, each of which has quite well-defined objectives. Phase I trials aim to establish relevant dose levels and toxicities, Phase II trials focus on a narrowly selected group of patients expected to show the most dramatic response, and Phase III trials are a full investigation of the treatment effects on patients broadly representative of the clinical population.

In investigations with some technological relevance, even if there is not an immediate focus on a decision to be made, questions will arise as to the practical implications of the conclusions. Is a difference established big enough to be of public health relevance in an epidemiological context, of relevance to farmers in an agricultural context or of engineering relevance in an industrial context? Do the conditions of the investigation justify extrapolation to the work-

A BROADER VIEW

ing context? To some extent such questions can be anticipated by appropriate design.

In both scientific and technological studies estimation of effects is likely to lead on to the further crucial question: what is the underlying process explaining what has been observed? Sometimes this is expressed via a search for causality. So far as possible these questions should be anticipated in design, especially in the definition of treatments and observations, but it is relatively rare for such explanations to be other than tentative and indeed they typically raise fresh issues for investigation.

It is sometimes argued that quite firm conclusions about causality are justified from experiments in which treatment allocation is made by objective randomization but not otherwise, it being particularly hazardous to draw causal conclusions from observational studies.

These issues are somewhat outside the scope of the present book but will be touched on in Section 2.5 after the discussion of the role of randomization. In the meantime some of the potential implications for design can be seen from the following Illustration.

Illustration. In an agricultural field trial a number of treatments are randomly assigned to plots, the response variable being the yield of product. One treatment, S, say, produces a spectacular growth of product, much higher than that from other treatments. The growth attracts birds from many kilometres around, the birds eat most of the product and as a result the final yield for S is very low. Has S caused a depression in yield?

The point of this illustration, which can be paralleled from other areas of application, is that the yield on the plots receiving S is indeed lower than the yield would have been on those plots had they been allocated to other treatments. In that sense, which meets one of the standard definitions of causality, allocation to S has thus caused a lowered yield. Yet in terms of understanding, and indeed practical application, that conclusion on its own is quite misleading. To understand the process leading to the final responses it is essential to observe and take account of the unanticipated intervention, the birds, which was supplementary to and dependent on the primary treatments. Preferably also intermediate variables should be recorded, for example, number of plants per square metre and measures of growth at various time points in the growing cycle. These will enable at least a tentative account to be developed of the process leading to the treatment differences in final yield which are the ultimate objective of study. In this way not only are treatment differences estimated but some partial understanding is built of the interpretation of such differences. This is a potentially causal explanation at a deeper level.

Such considerations may arise especially in situations in which a fairly long process intervenes between treatment allocation and the measurement of response.

These issues are quite pressing in some kinds of clinical trial, especially those in which patients are to be followed for an appreciable time. In the simplest case of randomization between two treatments, T and C, there is the possibility that some patients, called noncompliers, do not follow the regime to which they have been allocated. Even those who do comply may take supplementary medication and the tendency to do this may well be different in the two treatment groups. One approach to analysis, the so-called *intention-to-treat principle*, can be summarized in the slogan "ever randomized always analysed": one simply compares outcomes in the two treatment arms regardless of compliance or noncompliance. The argument, parallel to the argument in the agricultural example, is that if, say, patients receiving T do well, even if few of them comply with the treatment regimen, then the consequences of allocation to T are indeed beneficial, even if not necessarily because of the direct consequences of the treatment regimen.

Unless noncompliance is severe, the intention-to-treat analysis will be one important analysis but a further analysis taking account of any appreciable noncompliance seems very desirable. Such an analysis will, however, have some of the features of an observational study and the relatively clearcut conclusions of the analysis of a fully compliant study will be lost to some extent at least.

1.9 Bibliographic notes

While many of the ideas of experimental design have a long history, the first major systematic discussion was by R. A. Fisher (1926) in the context of agricultural field trials, subsequently developed into his magisterial book (Fisher, 1935 and subsequent editions). Yates in a series of major papers developed the subject much further; see especially Yates (1935, 1936, 1937). Applications were initially largely in agriculture and the biological sciences and then subsequently in industry. The paper by Box and Wilson (1951) was

FURTHER RESULTS AND EXERCISES   15

particularly influential in an industrial context. Recent industrial applications have been particularly associated with the name of the Japanese engineer, G. Taguchi. General books on scientific research that include some discussion of experimental design include Wilson (1952) and Beveridge (1952).

Of books on the subject, Cox (1958) emphasizes general principles in a qualitative discussion, Box, Hunter and Hunter (1978) emphasize industrial experiments and Hinkelman and Kempthorne (1994), a development of Kempthorne (1952), is closer to the originating agricultural applications. Piantadosi (1997) gives a thorough account of the design and analysis of clinical trials.

Vajda (1967a, 1967b) and Street and Street (1987) emphasize the combinatorial problems of design construction. Many general books on statistical methods have some discussion of design but tend to put their main emphasis on analysis; see especially Montgomery (1997). For very careful and systematic expositions with some emphasis respectively on industrial and biometric applications, see Dean and Voss (1999) and Clarke and Kempson (1997).

An annotated bibliography of papers up to the late 1960's is given by Herzberg and Cox (1969).

The notion of causality has a very long history although traditionally from a nonprobabilistic viewpoint. For accounts with a statistical focus, see Rubin (1974), Holland (1986), Cox (1992) and Cox and Wermuth (1996; section 8.7). Rather different views of causality are given by Dawid (2000), Lauritzen (2000) and Pearl (2000). For a discussion of compliance in clinical trials, see the papers edited by Goetghebeur and van Houwelingen (1998).

New mathematical developments in the design of experiments may be found in the main theoretical journals. More applied papers may also contain ideas of broad interest. For work with a primarily industrial focus, see *Technometrics*, for general biometric material, see *Biometrics*, for agricultural issues see the *Journal of Agricultural Science* and for specialized discussion connected with clinical trials see *Controlled Clinical Trials, Biostatistics* and *Statistics in Medicine*. *Applied Statistics* contains papers with a wide range of applications.

1.10 Further results and exercises

1. A study of the association between car telephone usage and accidents was reported by Redelmeier and Tibshirani (1997a)

and a further careful account discussed in detail the study design (Redelmeier and Tibshirani, 1997b). A randomized trial was infeasible on ethical grounds, and the investigators decided to conduct a case-control study. The cases were those individuals who had been in an automobile collision involving property damage (but not personal injury), who owned car phones, and who consented to having their car phone usage records reviewed.

(a) What considerations would be involved in finding a suitable control for each case?

(b) The investigators decided to use each case as his own control, in a specialized version of a case-control study called a case-crossover study. A "case driving period" was defined to be the ten minutes immediately preceding the collision. What considerations would be involved in determining the control period?

(c) An earlier study compared the accident rates of a group of drivers who owned cellular telephones to a group of drivers who did not, and found lower accident rates in the first group. What potential biases could affect this comparison?

2. A prospective case-crossover experiment to investigate the effect of alcohol on blood œstradiol levels was reported by Ginsberg et al. (1996). Two groups of twelve healthy postmenopausal women were investigated. One group was regularly taking œstrogen replacement therapy and the second was not. On the first day half the women in each group drank an alcoholic cocktail, and the remaining women had a similar juice drink without alcohol. On the second day the women who first had alcohol were given the plain juice drink and vice versa. In this manner it was intended that each woman serve as her own control.

(a) What precautions might well have been advisable in such a context to avoid bias?

(b) What features of an observational study does this study have?

(c) What features of an experiment does this study have?

3. Find out details of one or more medical studies the conclusions from which have been reported in the press recently. Were they experiments or observational studies? Is the design (or analysis) open to serious criticism?

FURTHER RESULTS AND EXERCISES   17

4. In an experiment to compare a number of alternative ways of treating back pain, pain levels are to be assessed before and after a period of intensive treatment. Think of a number of ways in which pain levels might be measured and discuss their relative merits. What measurements other than pain levels might be advisable?

5. As part of a study of the accuracy and precision of laboratory chemical assays, laboratories are provided with a number of nominally identical specimens for analysis. They are asked to divide each specimen into two parts and to report the separate analyses. Would this provide an adequate measure of reproducibility? If not recommend a better procedure.

6. Some years ago there was intense interest in the possibility that cloud-seeding by aircraft depositing silver iodide crystals on suitable cloud would induce rain. Discuss some of the issues likely to arise in studying the effect of cloud-seeding.

7. Preece et al. (1999) simulated the effect of mobile phone signals on cognitive function as follows. Subjects wore a headset and were subject to (i) no signal, (ii) a 915 MHz sine wave analogue signal, (iii) a 915 MHz sine wave modulated by a 217 Hz square wave. There were 18 subjects, and each of the six possible orders of the three conditions were used three times. After two practice sessions the three experimental conditions were used for each subject with 48 hours between tests. During each session a variety of computerized tests of mental efficiency were administered. The main result was that a particular reaction time was shorter under the condition (iii) than under (i) and (ii) but that for 14 other types of measurement there were no clear differences. Discuss the appropriateness of the control treatments and the extent to which stability of treatment differences across sessions might be examined.

8. Consider the circumstances under which the use of two different control groups might be valuable. For discussion of this for observational studies, where the idea is more commonly used, see Rosenbaum (1987).

CHAPTER 2

Avoidance of bias

2.1 General remarks

In Section 1.4 we stated a primary objective in the design of experiments to be the avoidance of bias, or systematic error. There are essentially two ways to reduce the possibility of bias. One is the use of randomization and the other the use in analysis of retrospective adjustments for perceived sources of bias. In this chapter we discuss randomization and retrospective adjustment in detail, concentrating to begin with on a simple experiment to compare two treatments T and C. Although bias removal is a primary objective of randomization, we discuss also its important role in giving estimates of errors of estimation.

2.2 Randomization

2.2.1 Allocation of treatments

Given $n = 2r$ experimental units we have to determine which are to receive T and which C. In most contexts it will be reasonable to require that the same numbers of units receive each treatment, so that the issue is how to allocate r units to T. Initially we suppose that there is no further information available about the units. Empirical evidence suggests that methods of allocation that involve ill-specified personal choices by the investigator are often subject to bias. This and, in particular, the need to establish publically independence from such biases suggest that a wholly impersonal method of allocation is desirable. Randomization is a very important way to achieve this: we choose r units at random out of the $2r$. It is of the essence that randomization means the use of an objective physical device; it does not mean that allocation is vaguely haphazard or even that it is done in a way that looks effectively random to the investigator.

Illustrations. One aspect of randomization is its use to conceal the treatment status of individuals. Thus in an examination of the reliability of laboratory measurements specimens could be sent for analysis of which some are from different individuals and others duplicate specimens from the same individual. Realistic assessment of precision would demand concealment of which were the duplicates and hidden randomization would achieve this.

The terminology "double-blind" is often used in published accounts of clinical trials. This usually means that the treatment status of each patient is concealed both from the patient and from the treating physician. In a triple-blind trial it would be aimed to conceal the treatment status as well from the individual assessing the end-point response.

There are a number of ways that randomization can be achieved in a simple experiment with just two treatments. Suppose initially that the units available are numbered $U_1, \ldots, U_n$. Then all $(2r)!/(r!r!)$ possible samples of size r may in principle be listed and one chosen to receive T, giving each such sample equal chance of selection. Another possibility is that one unit may be chosen at random out of $U_1, \ldots, U_n$ to receive T, then a second out of the remainder and so on until r have been chosen, the remainder receiving C. Finally, the units may be numbered $1, \ldots, n$, a random permutation applied, and the first r units allocated say to T.

It is not hard to show that these three procedures are equivalent. Usually randomization is subject to certain balance constraints aimed for example to improve precision or interpretability, but the essential features are those illustrated here. This discussion assumes that the randomization is done in one step. If units are accrued into the experiment in sequence over time different procedures will be needed to achieve the same objective; see Section 2.4.

2.2.2 The assumption of unit-treatment additivity

We base our initial discussion on the following assumption that can be regarded as underlying also many of the more complex designs and analyses developed later. We state the assumption for a general problem with v treatments, $T_1, \ldots, T_v$, using the simpler notation T, C for the special case $v = 2$.

RANDOMIZATION

Assumption of unit-treatment additivity. There exist constants, ξ_s for $s = 1, \ldots, n$, one for each unit, and constants $\tau_j, j = 1, \ldots, v$, one for each treatment, such that if T_j is allocated to U_s the resulting response is

$$\xi_s + \tau_j, \tag{2.1}$$

regardless of the allocation of treatments to other units.

The assumption is based on a full specification of responses corresponding to any possible treatment allocated to any unit, i.e. for each unit v possible responses are postulated. Now only one of these can be observed, namely that for the treatment actually implemented on that unit. The other $v - 1$ responses are counterfactuals. Thus the assumption can be tested at most indirectly by examining some of its observable consequences.

An apparently serious limitation of the assumption is its deterministic character. That is, it asserts that the difference between the responses for any two treatments is exactly the same for all units. In fact all the consequences that we shall use follow from the more plausible extended version in which some variation in treatment effect is allowed.

Extended assumption of unit-treatment additivity. With otherwise the same notation and assumptions as before, we assume that the response if T_j is applied to U_s is

$$\xi_s + \tau_j + \eta_{js}, \tag{2.2}$$

where the η_{js} are independent and identically distributed random variables which may, without loss of generality, be taken to have zero mean.

Thus the treatment difference between any two treatments on any particular unit s is modified by addition of a random term which is the difference of the two random terms in the original specification.

The random terms η_{js} represent two sources of variation. The first is *technical error* and represents an error of measurement or sampling. To this extent its variance can be estimated if independent duplicate measurements or samples are taken. The second is real variation in treatment effect from unit to unit, or what will later be called treatment by unit interaction, and this cannot be estimated separately from variation among the units, i.e. the variation of the ξ_s.

For simplicity all the subsequent calculations using the assumption of unit-treatment additivity will be based on the simple version of the assumption, but it can be shown that the conclusions all hold under the extended form.

The assumption of unit-treatment additivity is not directly testable, as only one outcome can be observed on each experimental unit. However, it can be indirectly tested by examining some of its consequences. For example, it may be possible to group units according to some property, using supplementary information about the units. Then if the treatment effect is estimated separately for different groups the results should differ only by random errors of estimation.

The assumption of unit-treatment additivity depends on the particular form of response used, and is not invariant under nonlinear transformations of the response. For example the effect of treatments on a necessarily positive response might plausibly be multiplicative, suggesting, for some purposes, a log transformation. Unit-treatment additivity implies also that the variance of response is the same for all treatments, thus allowing some test of the assumption without further information about the units and, in some cases at least, allowing a suitable scale for response to be estimated from the data on the basis of achieving constant variance.

Under the assumption of unit-treatment additivity it is sometimes reasonable to call the difference between τ_{j_1} and τ_{j_2} the *causal* effect of T_{j_1} compared with T_{j_2}. It measures the difference, under the general conditions of the experiment, between the response under T_{j_1} and the response that would have been obtained under T_{j_2}.

In general, the formulation as $\xi_s + \tau_j$ is overparameterized and a constraint such as $\Sigma \tau_j = 0$ can be imposed without loss of generality. For two treatments it is more symmetrical to write the responses under T and C as respectively

$$\xi_s + \delta, \qquad \xi_s - \delta, \qquad (2.3)$$

so that the treatment difference of interest is $\Delta = 2\delta$.

The assumption that the response on one unit is unaffected by which treatment is applied to another unit needs especial consideration when physically the same material is used as an experimental unit more than once. We return to further discussion of this point in Section 4.3.

2.2.3 Equivalent linear model

The simplest model for the comparison of two treatments, T and C, in which all variation other than the difference between treatments is regarded as totally random, is a linear model of the following form. Represent the observations on T and on C by random variables

$$Y_{T1},\ldots,Y_{Tr}; \quad Y_{C1},\ldots,Y_{Cr} \qquad (2.4)$$

and suppose that

$$E(Y_{Tj}) = \mu + \delta, \quad E(Y_{Cj}) = \mu - \delta. \qquad (2.5)$$

This is merely a convenient reparameterization of a model assigning the two groups arbitrary expected values. Equivalently we write

$$Y_{Tj} = \mu + \delta + \epsilon_{Tj}, \quad Y_{Cj} = \mu - \delta + \epsilon_{Cj}, \qquad (2.6)$$

where the random variables ϵ have by definition zero expectation.

To complete the specification more has to be set out about the distribution of the ϵ. We identify two possibilities.

Second moment assumption. The ϵ_j are mutually uncorrelated and all have the same variance, σ^2.

Normal theory assumption. The ϵ_j are independently normally distributed with constant variance.

The least squares estimate of Δ, the difference of the two means, is

$$\hat{\Delta} = \bar{Y}_{T.} - \bar{Y}_{C.}, \qquad (2.7)$$

where $\bar{Y}_{T.}$ is the mean response on the units receiving T, and $\bar{Y}_{C.}$ is the mean response on the units receiving C. Here and throughout we denote summation over a subscript by a full stop. The residual mean square is

$$s^2 = \Sigma\{(Y_{Tj} - \bar{Y}_{T.})^2 + (Y_{Cj} - \bar{Y}_{C.})^2\}/(2r-2). \qquad (2.8)$$

Defining the estimated variance of $\hat{\Delta}$ by

$$\operatorname{evar}(\hat{\Delta}) = 2s^2/r, \qquad (2.9)$$

we have under (2.6) and the second moment assumptions that

$$E(\hat{\Delta}) = \Delta, \qquad (2.10)$$
$$E\{\operatorname{evar}(\hat{\Delta})\} = \operatorname{var}(\hat{\Delta}). \qquad (2.11)$$

The optimality properties of the estimates of Δ and $\operatorname{var}(\hat{\Delta})$ under both the second moment assumption and the normal theory assumption follow from the same results in the general linear model and are detailed in Appendix A. For example, under the second moment assumption $\hat{\Delta}$ is the minimum variance unbiased estimate that is linear in the observations, and under normality is the minimum variance estimate among all unbiased estimates. Of course, such optimality considerations, while reassuring, take no account of special concerns such as the presence of individual defective observations.

2.2.4 Randomization-based analysis

We now develop a conceptually different approach to the analysis assuming unit-treatment additivity and regarding probability as entering only via the randomization used in allocating treatments to the experimental units.

We again write random variables representing the observations on T and C respectively

$$Y_{T1}, \ldots, Y_{Tr}, \qquad (2.12)$$
$$Y_{C1}, \ldots, Y_{Cr}, \qquad (2.13)$$

where the order is that obtained by, say, the second scheme of randomization specified in Section 2.2.1. Thus Y_{T1}, for example, is equally likely to arise from any of the $n = 2r$ experimental units. With P_R denoting the probability measure induced over the experimental units by the randomization, we have that, for example,

$$\begin{aligned} P_R(Y_{Tj} \in U_s) &= (2r)^{-1}, \\ P_R(Y_{Tj} \in U_s, Y_{Ck} \in U_t, s \neq t) &= \{2r(2r-1)\}^{-1}, \end{aligned} \qquad (2.14)$$

where unit U_s is the jth to receive T.

Suppose now that we estimate both $\Delta = 2\delta$ and its standard error by the linear model formulae for the comparison of two independent samples, given in equations (2.7) and (2.9). The properties of these estimates under the probability distribution induced by the randomization can be obtained, and the central results are that in parallel to (2.10) and (2.11),

$$\begin{aligned} E_R(\hat{\Delta}) &= \Delta, & (2.15) \\ E_R\{\text{evar}(\hat{\Delta})\} &= \text{var}_R(\hat{\Delta}), & (2.16) \end{aligned}$$

where E_R and var_R denote expectation and variance calculated under the randomization distribution.

RANDOMIZATION

We may call these second moment properties of the randomization distribution. They are best understood by examining a simple special case, for instance $n = 2r = 4$, when the $4! = 24$ distinct permutations lead to six effectively different treatment arrangements.

The simplest proof of (2.15) and (2.16) is obtained by introducing indicator random variables with, for the sth unit, I_s taking values 1 or 0 according as T or C is allocated to that unit.

The contribution of the sth unit to the sample total for Y_T. is thus

$$I_s(\xi_s + \delta), \qquad (2.17)$$

whereas the contribution for C is

$$(1 - I_s)(\xi_s - \delta). \qquad (2.18)$$

Thus

$$\hat{\Delta} = \Sigma\{I_s(\xi_s + \delta) - (1 - I_s)(\xi_s - \delta)\}/r \qquad (2.19)$$

and the probability properties follow from those of I_s.

A more elegant and general argument is in outline as follows:

$$\bar{Y}_T. - \bar{Y}_C. = \Delta + L(\xi), \qquad (2.20)$$

where $L(\xi)$ is a linear combination of the ξ's depending on the particular allocation. Now $E_R(L)$ is a symmetric linear function, i.e. is invariant under permutation of the units. Therefore

$$E_R(L) = a\Sigma\xi_s, \qquad (2.21)$$

say. But if $\xi_s = \xi$ is constant for all s, then $L = 0$ which implies $a = 0$.

Similarly both $\text{var}_R(\hat{\Delta}), E_R\{\text{evar}(\hat{\Delta})\}$ do not depend on Δ and are symmetric second degree functions of $\xi_1, \ldots, \xi_{2r}$ vanishing if all ξ_s are equal. Hence

$$\text{var}_R(\hat{\Delta}) = b_1\Sigma(\xi_s - \bar{\xi}.)^2,$$
$$E_R\{\text{evar}(\hat{\Delta})\} = b_2\Sigma(\xi_s - \bar{\xi}.)^2,$$

where b_1, b_2 are constants depending only on n. To find the b's we may choose any special ξ's, such as $\xi_1 = 1, \xi_s = 0, (s \neq 1)$ or suppose that $\xi_1, \ldots, \xi_n$ are independent and identically distributed random variables with mean zero and variance ψ^2. This is a technical mathematical trick, not a physical assumption about the variability.

Let $\mathcal{E}$ denote expectation with respect to that distribution and apply $\mathcal{E}$ to both sides of last two equations. The expectations on

the left are known and
$$\mathcal{E}\Sigma(\xi_s - \bar{\xi}_.)^2 = (2r-1)\psi^2; \qquad (2.22)$$
it follows that
$$b_1 = b_2 = 2/\{r(2r-1)\}. \qquad (2.23)$$

Thus standard two-sample analysis based on an assumption of independent and identically distributed errors has a second moment justification under randomization theory via unit-treatment additivity. The same holds very generally for designs considered in later chapters.

The second moment optimality of these procedures follows under randomization theory in essentially the same way as under a physical model. There is no obvious stronger optimality property solely in a randomization-based framework.

2.2.5 Randomization test and confidence limits

Of more direct interest than $\hat{\Delta}$ and $\mathrm{evar}(\hat{\Delta})$ is the pivotal statistic
$$(\hat{\Delta} - \Delta)/\sqrt{\mathrm{evar}(\hat{\Delta})} \qquad (2.24)$$
that would generate confidence limits for Δ but more complicated arguments are needed for direct analytical examination of its randomization distribution.

Although we do not in this book put much emphasis on tests of significance we note briefly that randomization generates a formally exact test of significance and confidence limits for Δ. To see whether Δ_0 is in the confidence region at a given level we subtract Δ_0 from all values in T and test $\Delta = 0$.

This null hypothesis asserts that the observations are totally unaffected by treatment allocation. We may thus write down the observations that would have been obtained under all possible allocations of treatments to units. Each such arrangement has equal probability under the null hypothesis. The distribution of any test statistic then follows. Using the constraints of the randomization formulation, simplification of the test statistic is often possible.

We illustrate these points briefly on the comparison of two treatments T and C, with equal numbers of units for each treatment and randomization by one of the methods of Section 2.2.2. Suppose that the observations are
$$\mathcal{P} = \{y_{T1}, \ldots, y_{Tr}; y_{C1}, \ldots, y_{Cr}\}, \qquad (2.25)$$

RANDOMIZATION   27

which can be regarded as forming a finite population $\mathcal{P}$. Write $m_\mathcal{P}, w_\mathcal{P}$ for the mean and effective variance of this finite population defined as

$$m_\mathcal{P} = \Sigma y_u/(2r), \tag{2.26}$$
$$w_\mathcal{P} = \Sigma(y_u - m_\mathcal{P})^2/(2r-1), \tag{2.27}$$

where the sum is over all members of the finite population. To test the null hypothesis a test statistic has to be chosen that is defined for every possible treatment allocation. One natural choice is the two-sample Student t statistic. It is easily shown that this is a function of the constants $m_\mathcal{P}, w_\mathcal{P}$ and of $\bar{Y}_{T.}$, the mean response of the units receiving T. Only $\bar{Y}_{T.}$ is a random variable over the various treatment allocations and therefore we can treat it as the test statistic.

It is possible to find the exact distribution of $\bar{Y}_{T.}$ under the null hypothesis by enumerating all distinct samples of size r from $\mathcal{P}$ under sampling without replacement. Then the probability of a value as or more extreme than the observed value $\bar{y}_{T.}$ can be found. Alternatively we may use the theory of sampling without replacement from a finite population to show that

$$E_R(\bar{Y}_{T.}) = m_\mathcal{P}, \tag{2.28}$$
$$\text{var}_R(\bar{Y}_{T.}) = w_\mathcal{P}/(2r). \tag{2.29}$$

Higher moments are available but in many contexts a strong central limit effect operates and a test based on a normal approximation for the null distribution of $\bar{Y}_{T.}$ will be adequate.

A totally artificial illustration of these formulae is as follows. Suppose that $r = 2$ and that the observations on T and C are respectively 3, 1 and $-1, -3$. Under the null hypothesis the possible values of observations on T corresponding to the six choices of units to be allocated to T are

$$(-1,-3); (-1,1); (-1,3); (-3,1); (-3,3); (1,3) \tag{2.30}$$

so that the induced randomization distribution of $\bar{Y}_{T.}$ has mass $1/6$ at $-2, -1, 1, 2$ and mass $1/3$ at 0. The one-sided level of significance of the data is $1/6$. The mean and variance of the distribution are respectively 0 and $5/3$; note that the normal approximation to the significance level is $\Phi(-2\sqrt{3}/\sqrt{5}) \simeq 0.06$, which, considering the extreme discreteness of the permutation distribution, is not too far from the exact value.

2.2.6 More than two treatments

The previous discussion has concentrated for simplicity on the comparison of two treatments T and C. Suppose now that there are v treatments $T_1, \ldots, T_v$. In many ways the previous discussion carries through with little change.

The first new point of design concerns whether the same number of units should be assigned to each treatment. If there is no obvious structure to the treatments, so that for instance all comparisons of pairs of treatments are of equal interest, then equal replication will be natural and optimal, for example in the sense of minimizing the average variance over all comparisons of pairs of treatments. Unequal interest in different comparisons may suggest unequal replication.

For example, suppose that there is a special treatment T_0, possibly a control, and v ordinary treatments and that particular interest focuses on the comparisons of the ordinary treatments with T_0. Suppose that each ordinary treatment is replicated r times and that T_0 occurs cr times. Then the variance of a difference of interest is proportional to

$$1/r + 1/(cr) \tag{2.31}$$

and we aim to minimize this subject to a given total number of observations $n = r(v + c)$. We eliminate r and obtain a simple approximation by regarding c as a continuous variable; the minimum is at $c = \sqrt{v}$. With three or four ordinary treatments there is an appreciable gain in efficiency, by this criterion, by replicating T_0 up to twice as often as the other treatments.

The assumption of unit-treatment additivity is as given at (2.1). The equivalent linear model is

$$Y_{js} = \mu + \tau_j + \epsilon_{js} \tag{2.32}$$

with $j = 1, \ldots, v$ and $s = 1, \ldots, r$. An important aspect of having more than two treatments is that we may be interested in more complicated comparisons than simple differences. We shall call $\Sigma l_j \tau_j$, where $\Sigma l_j = 0$, a *treatment contrast*. The special case of a difference $\tau_{j_1} - \tau_{j_2}$ is called a *simple contrast* and examples of more general contrasts are

$$(\tau_1 + \tau_2 + \tau_3)/3 - \tau_5, \qquad (\tau_1 + \tau_2 + \tau_3)/3 - (\tau_4 + \tau_6)/2. \tag{2.33}$$

We defer more detailed discussion of contrasts to Section 3.5.2 but in the meantime note that the general contrast $\Sigma l_j \tau_j$ is estimated by $\Sigma l_j \bar{Y}_{j.}$ with, in the simple case of equal replication, variance

$$\Sigma l_j^2 \sigma^2 / r, \tag{2.34}$$

estimated by replacing σ^2 by the mean square within treatment groups. Under complete randomization and the assumption of unit-treatment additivity the correspondence between the properties found under the physical model and under randomization theory discussed in Section 2.2.4 carries through.

2.3 Retrospective adjustment for bias

Even with carefully designed experiments there may be a need in the analysis to make some adjustment for bias. In some situations where randomization has been used, there may be some suggestion from the data that either by accident effective balance of important features has not been achieved or that possibly the implementation of the randomization has been ineffective. Alternatively it may not be practicable to use randomization to eliminate systematic error.

Sometimes, especially with well-standardized physical measurements, such corrections are done on an *a priori* basis.

Illustration. It may not be feasible precisely to control the temperature at which the measurement on each unit is made but the temperature dependence of the property in question, for example electrical resistance, may be known with sufficient precision for an *a priori* correction to be made.

For the remainder of the discussion, we assume that any bias correction has to be estimated internally from the data.

In general we suppose that on the sth experimental unit there is available a $q \times 1$ vector z_s of baseline explanatory variables, measured in principle before randomization. For simplicity we discuss mostly $q = 1$ and two treatment groups.

If the relevance of z is recognized at the design stage then the completely random assignment of treatments to units discussed in this chapter is inappropriate unless there are practical constraints that prevent the randomization scheme being modified to take account of z. We discuss such randomization schemes in Chapters 3 and 4. If, however, the relevance of z is only recognized retrospectively, it will be important to check that it is indeed properly regarded as a baseline variable and, if there is a serious lack of balance between the treatment groups with respect to z, to consider whether there is a reasonable explanation as to why this difference occurred in a context where randomization should, with high probability, have removed any substantial bias.

Suppose, however, that an apparent bias does exist. Figure 2.1 shows three of the various possibilities that can arise. In Fig. 2.1a the clear lack of parallelism means that a single estimate of treatment difference is at best incomplete and will depend on the particular value of z used for comparison. Alternatively a transformation of the response scale inducing parallelism may be found. In Fig. 2.1b the crossing over in the range of the data accentuates the dangers of a single comparison; the qualitative consequences may well be stronger than those in the situation of Fig. 2.1a. Finally in Fig. 2.1c the effective parallelism of the two relations suggests a correction for bias equivalent to using the vertical difference between the two lines as an estimated treatment difference preferable to a direct comparison of unadjusted means which in the particular instance shown underestimates the difference between T and C.

A formulation based on a linear model is to write

$$E(Y_{Tj}) = \mu + \delta + \beta(z_{Tj} - \bar{z}_{..}), \qquad (2.35)$$
$$E(Y_{Cj}) = \mu - \delta + \beta(z_{Cj} - \bar{z}_{..}), \qquad (2.36)$$

where z_{Tj} and z_{Cj} are the values of z on the jth units to receive T and C, respectively, and $\bar{z}_{..}$ the overall average z. The inclusion of $\bar{z}_{..}$ is not essential but preserves an interpretation for μ as the expected value of $\bar{Y}_{...}$.

We make the normal theory or second moment assumptions about the error terms as in Section 2.2.3. Note that $\Delta = 2\delta$ measures the difference between T and C in the expected response at any fixed value of z. Provided that z is a genuine baseline variable, and the assumption of parallelism is satisfied, Δ remains a measure of the effect of changing the treatment from C to T.

If z is a $q \times 1$ vector the only change is to replace βz by $\beta^T z$, β becoming a $q \times 1$ vector of parameters; see also Section 3.6.

A least squares analysis of the model gives for scalar z,

$$\begin{pmatrix} 2r & 0 & 0 \\ 0 & 2r & r(\bar{z}_{T.} - \bar{z}_{C.}) \\ 0 & r(\bar{z}_{T.} - \bar{z}_{C.}) & S_{zz} \end{pmatrix} \begin{pmatrix} \hat{\mu} \\ \hat{\delta} \\ \hat{\beta} \end{pmatrix}$$
$$= \begin{pmatrix} r(\bar{Y}_{T.} + \bar{Y}_{C.}) \\ r(\bar{Y}_{T.} - \bar{Y}_{C.}) \\ \Sigma\{Y_{Tj}(z_{Tj} - \bar{z}_{..}) + Y_{Cj}(z_{Cj} - \bar{z}_{..})\} \end{pmatrix}, \qquad (2.37)$$

RETROSPECTIVE ADJUSTMENT FOR BIAS

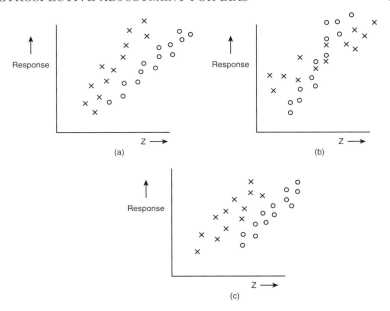

Figure 2.1 *Two treatments:* ×, *T and* ∘, *C. Response variable, Y; Baseline variable, z measured before randomization and therefore unaffected by treatments. In (a) nonparallelism means that unadjusted estimate of treatment effect is biased, and adjustment depends on particular values of z. In (b) crossing over of relations means that even qualitative interpretation of treatment effect is different at different values of z. In (c) essential parallelism means that treatment effect can be estimated from vertical displacement of lines.*

where
$$S_{zz} = \Sigma\{(z_{Tj} - \bar{z}_{..})^2 + (z_{Cj} - \bar{z}_{..}^2)\}. \tag{2.38}$$
The least squares equations yield in particular
$$\hat{\Delta} = 2\hat{\delta} = \bar{Y}_{T.} - \bar{Y}_{C.} - \hat{\beta}(\bar{z}_{T.} - \bar{z}_{C.}), \tag{2.39}$$
where $\hat{\beta}$ is the least squares estimated slope, agreeing precisely with the informal geometric argument given above. It follows also either by a direct calculation or by inverting the matrix of the least squares equations that
$$\text{var}(\hat{\Delta}) = \sigma^2\{2/r + (\bar{z}_{T.} - \bar{z}_{C.})^2/R_{zz}\}, \tag{2.40}$$
where
$$R_{zz} = \Sigma\{(z_{Tj} - \bar{z}_{T.})^2 + (z_{Cj} - \bar{z}_{C.})^2\}. \tag{2.41}$$

For a given value of σ^2 the variance is inflated as compared with that for the difference between two means because of sampling error in $\hat{\beta}$.

This procedure is known in the older literature as *analysis of covariance*. If a standard least squares regression package is used to do the calculations it may be simpler to use more direct parameterizations than that used here although the advantages of choosing parameters which have a direct interpretation should not be overlooked. In extreme cases the subtraction of a suitable constant, not necessarily the overall mean, from z, and possibly rescaling by a power of 10, may be needed to avoid numerical instability and also to aid interpretability.

We have presented the bias correction via an assumed linear model. The relation with randomization theory does, however, need discussion: have we totally abandoned a randomization viewpoint? First, as we have stressed, if the relevance of the bias inducing variable z had been clear from the start then normally this bias would have been avoided by using a different form of randomization, for example a randomized block design; see Chapter 3. When complete randomization has, however, been used and the role of z is considered retrospectively then the quantity $\bar{z}_{T.} - \bar{z}_{C.}$, which is a random variable under randomization, becomes an ancillary statistic. That is, to be relevant to the inference under discussion the ensemble of hypothetical repetitions should hold $\bar{z}_{T.} - \bar{z}_{C.}$ fixed, either exactly or approximately. It is possible to hold this ancillary exactly fixed only in special cases, notably when z corresponds to qualitative groupings of the units. Otherwise it can be shown that an appropriate notion of approximate conditioning induces the appropriate randomization properties for the analysis of covariance estimate of Δ, and in that sense there is no conflict between randomization theory and that based on an assumed linear model. Put differently, randomization approximately justifies the assumed linear model.

2.4 Some more on randomization

In theory randomization is a powerful notion with a number of important features.

First it removes bias.

Secondly it allows what can fairly reasonably be called causal inference in that if a clear difference between two treatment groups arises it can only be either an accident of the randomization or a consequence of the treatment.

Thirdly it allows the calculation of an estimated variance for a treatment contrast in the above and many other situations based only on a single assumption of unit-treatment additivity and without the need to supply an *ad hoc* model for each new design.

Finally it allows the calculation of confidence limits for treatment differences, in principle based on unit-treatment additivity alone.

In practice the role of randomization ranges from being crucial in some contexts to being of relatively minor importance in others; we do stress, however, the general desirability of impersonal allocation schemes.

There are, moreover, some conceptual difficulties when we consider more realistic situations. The discussion so far, except for Section 2.3, has supposed both that there is no baseline information available on the experimental units and that the randomization is in effect done in one operation rather than sequentially in time.

The absence of baseline information means that all arrangements of treatments can be regarded on an equal footing in the randomization-induced probability calculations. In practice potentially relevant information on the units is nearly always available. The broad strategy of the subsequent chapters is that such information, when judged important, is taken account of in the design, in particular to improve precision, and randomization used to safeguard against other sources of variation. In the last analysis some set of designs is regarded as on an equal footing to provide a relevant reference set for inference. Additional features can in principle be covered by the adjustments of the type discussed in Section 2.3, but some frugality in the use of this idea is needed, especially where there are many baseline variables.

A need to perform the randomization one unit at a time, such as in clinical trials in which patients are accrued in sequence over an appreciable time, raises different issues unless decisions about different patients are quite separate, for example by virtually always being in different centres. For example, if a single group of $2r$ patients is to be allocated equally to two treatments and this is done sequentially, a point will almost always be reached where all individuals must be allocated to a particular treatment in order to force the necessary balance and the advantages of concealment associated with the randomization are lost. On the other hand if all patients were independently randomized, there would be some chance, even if only a fairly small one, of extreme imbalance in numbers. The most reasonable compromise is to group the patients into successive blocks and to randomize ensuring balance of numbers within each block. A suitable number of patients per block is often between 6 and 12 thus in the first case ensuring that each block has three occurrences of each of two treatments. In line with the discussion in the next chapter it would often be reasonable to stratify by one or two important features; for example there might be separate blocks of men and women. Randomization schemes that adapt to the information available in earlier stages of the experiment are discussed in Section 8.2.

2.5 More on causality

We return to the issue of causality introduced in Section 1.8. For ease of exposition we again suppose there to be just two treatments, T and C, possibly a new treatment and a control. The counterfactual definition of causality introduced in Section 1.8, the notion that an individual receiving, say, T gives a response systematically different from the response that would have resulted had the individual received C, other things being equal, is encapsulated in the assumption of unit-treatment additivity in either its simple or in its extended form. Indeed the general notion may be regarded as an extension of unit-treatment additivity to possibly observational contexts.

In the above sense, causality can be inferred from a randomized experiment with uncertainty expressed via a significance test, for example via the randomization-based test of Section 2.2.5. The argument is direct. Suppose a set of experimental units is randomized between T and C, a response is observed, and a significance test shows very strong evidence against the null hypothesis of treatment identity and evidence, say, that the parameter Δ is positive. Then either an extreme chance fluctuation has occurred, to which the significance level of the test refers, or units receiving T have a higher response than they would have yielded had they received C and this is precisely the definition of causality under discussion.

The situation is represented graphically in Fig. 2.2. Randomization breaks the possible edge between the unobserved confounder U and treatment.

In a comparable observational study the possibility of an unobserved confounder affecting both treatment and response in a

MORE ON CAUSALITY   35

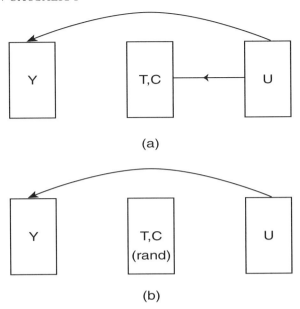

Figure 2.2 *Unobserved confounder, U; treatment, T, C; response, Y. No treatment difference, no edge between T, C and Y. In an observational study, (a), there are edges between U and other nodes. Marginalization over U can be shown to induce dependency between T, C and Y. In a randomized experiment, (b), randomization of T, C ensures there is no edge to it from U. Marginalization over U does not induce an edge between T, C and Y.*

systematic way would be an additional source of uncertainty, sometimes a very serious one, that would make any causal interpretation much more tentative.

This conclusion highlighting the advantage of randomized experiments over observational studies is very important. Nevertheless there are some qualifications to it.

First we have assumed the issues of noncompliance discussed in Section 1.8 are unimportant: the treatments as implemented are assumed to be genuinely those that it is required to study. An implication for design concerns the importance of measuring any features arising throughout the implementation of the experiment that might have an unanticipated distortion of the treatments from those that were originally specified.

Next it is assumed that randomization has addressed all sources of potential systematic error including any associated directly with the measurement of response.

The most important assumption, however, is that the treatment effect, Δ, is essentially constant and in particular does not have systematic sign reversals between different units. That is, in the terminology to be introduced later there is no major interaction between the treatment effect and intrinsic features of the experimental units.

In an extension of model (2.3) in which each experimental unit has its own treatment parameter, the difference estimated in a randomized experiment is the average treatment effect over the full set of units used in the experiment. If, moreover, these were a random sample from a population of units then the average treatment effect over that population is estimated. Such conclusions have much less force whenever the units used in the experiment are unrepresentative of some target population or if substantial and interpretable interactions occur with features of the experimental units.

There is a connection of these matters with the apparently antithetical notions of generalizability and specificity. Suppose for example that a randomized experiment shows a clear superiority in some sense of T over C. Under what circumstances may we reasonably expect the superiority to be reproduced over a new somewhat different set of units perhaps in different external conditions? This a matter of generalizability. On the other hand the question of whether T will give an improved response for a particular new experimental unit is one of specificity. Key aids in both aspects are understanding of underlying process and of the nature of any interactions of the treatment effect with features of the experimental units. Both these, and especially the latter, may help clarify the conditions under which the superiority of T may not be achieved. The main implication in the context of the present book concerns the importance of factorial experiments, to be discussed in Chapters 5 and 6, and in particular factorial experiments in which one or more of the factors correspond to properties of the experimental units.

2.6 Bibliographic notes

Formal randomization was introduced into the design of experiments by R. A. Fisher. The developments for agricultural experiments especially by Yates, as for example in Yates (1937), put cen-

FURTHER RESULTS AND EXERCISES

tral importance on achieving meaningful estimates of error via the randomization rather than via physical assumptions about the error structure. In some countries, however, this view has not gained wide acceptance. Yates (1951a,b) discussed randomization more systematically. For a general mathematical discussion of the basis of randomization theory, see Bailey and Rowley (1987). For a combinatorial nonprobabilistic formulation of the notion of randomization, see Singer and Pincus (1998).

Models based on unit-treatment additivity stem from Neyman (1923).

The relation between tests based on randomization and those stemming from normal theory assumptions was discussed in detail in early work by Welch (1937) and Pitman (1937). See Hinkelman and Kempthorne (1994) and Kempthorne (1952) for an account regarding the randomization analysis as primary. Manly (1997) emphasizes the direct role of randomization analyses in applications. For a discussion of the Central Limit Theorem, and Edgeworth and saddle-point expansions connected with sampling without replacement from a finite population, see Thompson (1997).

A priori corrections for bias are widely used, for example in the physical sciences for adjustments to standard temperature, etc. Corrections based explicitly on least squares analysis were the motivation for the development of analysis of covariance. For a review of analysis of covariance, see Cox and McCullagh (1982). Similar adjustments are central to the careful analysis of observational data to attempt to adjust for unwanted lack of comparability of groups. See, for example, Rosenbaum (1999) and references therein.

For references on causality, see the Bibliographic notes to Chapter 1.

2.7 Further results and exercises

1. Suppose that in the comparison of two treatments with r units for each treatment the observations are completely separated, for example that all the observations on T exceed all those on C. Show that the one-sided significance level under the randomization distribution is $(r!)^2/(2r)!$. Comment on the reasonableness or otherwise of the property that it does not depend on the numerical values and in particular on the distance apart of the two sets of observations.

2. In the comparison of v equally replicated treatments in a completely randomized design show that under a null hypothesis of no treatment effects the randomization expectation of the mean squares between and within treatments, defined in the standard way, are the same. What further calculations would be desirable to examine the distribution under randomization of the standard F statistic?

3. Suppose that on each unit a property, for example blood pressure, is measured before randomization and then the same property measured as a response after treatment. Discuss the relative merits of taking as response on each individual the difference between the values after and before randomization versus taking as response the measure after randomization and adjusting for regression on the value before. See Cox (1957, 1958, Chapter 4) and Cox and Snell (1981, Example D).

4. Develop the analysis first for two treatments and then for v treatments for testing the parallelism of the regression lines involved in a regression adjustment. Sketch some possible approaches to interpretation if nonparallelism is found.

5. Show that in the randomization analysis of the comparison of two treatments with a binary response, the randomization test of a null hypothesis of no effect is the exact most powerful conditional test of the equality of binomial parameters, usually called Fisher's exact test (Pearson, 1947; Cox and Hinkley, 1974, Chapter 5). If the responses are individually binomial, corresponding to the numbers of successes in, say, t trials show that a randomization test is essentially the standard Mantel-Haenszel test with a sandwich estimate of variance (McCullagh and Nelder, 1989, Chapter 14).

6. Discuss a randomization formulation of the situation of Exercise 5 in the nonnull case. See Copas (1973).

7. Suppose that in an experiment to compare two treatments, T and C, the response Y of interest is very expensive to measure. It is, however, relatively inexpensive to measure a surrogate response variable, X, thought to be quite highly correlated with Y. It is therefore proposed to measure X on all units and both X and Y on a subsample. Discuss some of the issues of design and analysis that this raises.

FURTHER RESULTS AND EXERCISES 39

8. Individual potential experimental units are grouped into clusters each of k individuals. A number of treatments are then randomized to clusters, i.e. all individuals in the same cluster receive the same treatment. What would be the likely consequences of analysing such an experiment as if the treatments had been randomized to individuals? Cornfield (1978) in the context of clinical trials called such an analysis "an exercise in self-deception". Was he justified?

9. Show that in a large completely randomized experiment under the model of unit-treatment additivity the sample cumulative distribution functions of response to the different treatments differ only by translations. How could such a hypothesis be tested nonparametrically? Discuss why in practice examination of homogeneity of variance would often be preferable. First for two treatments and then for more than two treatments suggest parametric and nonparametric methods for finding a monotone transformation inducing translation structure and for testing whether such a transformation exists. Nonparametric analysis of completely randomized and randomized block designs is discussed in Lehmann (1975).

10. Studies in various medical fields, for example psychiatry (Johnson, 1998), have shown that where the same treatment contrasts have been estimated both via randomized clinical trials and via observational studies, the former tend to show smaller advantages of new procedures than the latter. Why might this be?

11. When the sequential blocked randomization scheme of Section 2.4 is used in clinical trials it is relatively common to disregard the blocking in the statistical analysis. How might some justification be given of the disregard of the principle that constraints used in design should be reflected in the statistical model?

CHAPTER 3

Control of haphazard variation

3.1 General remarks

In the previous chapter the primary emphasis was on the elimination of systematic error. We now turn to the control of haphazard error, which may enter at any of the phases of an investigation. Sources of haphazard error include intrinsic variation in the experimental units, variation introduced in the intermediate phases of an investigation and measurement or sampling error in recording response.

It is important that measures to control the effect of such variation cover all the main sources of variation and some knowledge, even if rather qualitative, of the relative importance of the different sources is needed.

The ways in which the effect of haphazard variability can be reduced include the following approaches.

1. It may be possible to use more uniform material, improved measuring techniques and more internal replication, i.e. repeat observations on each unit.

2. It may be possible to use more experimental units.

3. The technique of blocking, discussed in detail below, is a widely applicable technique for improving precision.

4. Adjustment for baseline features by the techniques for bias removal discussed in Section 2.3 can be used.

5. Special models of error structure may be constructed, for example based on a time series or spatial model.

On the first two points we make here only incidental comments.

There will usually be limits to the increase in precision achievable by use of more uniform material and in technological experiments the wide applicability of the conclusions may be prejudiced if artificial uniformity is forced.

Illustration. In some contexts it may be possible to use pairs of homozygotic twins as experimental units in the way set out in detail in Section 3.3. There may, however, be some doubt as to whether conclusions apply to a wider population of individuals. More broadly, in a study to elucidate some new phenomenon or suspected effect it will usually be best to begin with the circumstances under which that effect occurs in its most clear-cut form. In a study in which practical application is of fairly direct concern the representativeness of the experimental conditions merits more emphasis, especially if it is suspected that the treatment effects have different signs in different individuals.

In principle precision can always be improved by increasing the number of experimental units. The standard error of treatment comparisons is inversely proportional to the square root of the number of units, provided the residual standard deviation remains constant. In practice the investigator's control may be weaker in large investigations than in small so that the theoretical increase in the number of units needed to shorten the resulting confidence limits for treatment effects is often an underestimate.

3.2 Precision improvement by blocking

The central idea behind blocking is an entirely commonsense one of aiming to compare like with like. Using whatever prior knowledge is available about which baseline features of the units and other aspects of the experimental set-up are strongly associated with potential response, we group the units into blocks such that all the units in any one block are likely to give similar responses in the absence of treatment differences. Then, in the simplest case, by allocating one unit in each block to each treatment, treatments are compared on units within the same block.

The formation of blocks is usually, however, quite constrained in addition by the way in which the experiment is conducted. For example, in a laboratory experiment a block might correspond to the work that can be done in a day. In our initial discussion we regard the different blocks as merely convenient groupings without individual interpretation. Thus it makes no sense to try to interpret differences between blocks, except possibly as a guide for future experimentation to see whether the blocking has been effective in error control. Sometimes, however, some aspects of blocking do have a clear interpretation, and then the issues of Chapter 5 concerned

MATCHED PAIRS

with factorial experiments apply. In such cases it is preferable to use the term *stratification* rather than blocking.

Illustrations. Typical ways of forming blocks are to group together neighbouring plots of ground, responses from one subject in one session of a psychological experiment under different conditions, batches of material produced on one machine, where several similar machines are producing nominally the same product, groups of genetically similar animals of the same gender and initial body weight, pairs of homozygotic twins, the two eyes of the same subject in an opthalmological experiment, and so on. Note, however, that if gender were a defining variable for blocks, i.e. strata, we would likely want not only to compare treatments but also to examine whether treatment differences are the same for males and females and this brings in aspects that we ignore in the present chapter.

3.3 Matched pairs

3.3.1 Model and analysis

Suppose that we have just two treatments, T and C, for comparison and that we can group the experimental units into pairs, so that in the absence of treatment differences similar responses are to be expected in the two units within the same pair or block.

It is now reasonable from many viewpoints to assign one member of the pair to T and one to C and, moreover, in the absence of additional structure, to randomize the allocation within each pair independently from pair to pair. This yields what we call the *matched pair* design.

Thus if we label the units

$$U_{11}, U_{21}; \quad U_{12}, U_{22}; \quad \ldots; \quad U_{1r}, U_{2r} \tag{3.1}$$

a possible design would be

$$T, C; \quad C, T; \quad \ldots; \quad T, C. \tag{3.2}$$

As in Chapter 2, a linear model that directly corresponds with randomization theory can be constructed. The broad principle in setting up such a physical linear model is that randomization constraints forced by the design are represented by parameters in the linear model. Writing Y_{Ts}, Y_{Cs} for the observations on treatment and control for the sth pair, we have the model

$$Y_{Ts} = \mu + \beta_s + \delta + \epsilon_{Ts}, \quad Y_{Cs} = \mu + \beta_s - \delta + \epsilon_{Cs}, \qquad (3.3)$$

where the ϵ are random variables of mean zero. As in Section 2.2, either the normal theory or the second moment assumption about the errors may be made; the normal theory assumption leads to distributional results and strong optimality properties.

Model (3.3) is overparameterized, but this is often convenient to achieve a symmetrical formulation. The redundancy could be avoided here by, for example, setting μ to any arbitrary known value, such as zero.

A least squares analysis of this model can be done in several ways. The simplest, for this very special case, is to transform the Y_{Ts}, Y_{Cs} to sums, B_s and differences, D_s. Because this is proportional to an orthogonal transformation, the transformed observations are also uncorrelated and have constant variance. Further in the linear model for the new variables we have

$$E(B_s) = 2(\mu + \beta_s), \quad E(D_s) = 2\delta = \Delta. \qquad (3.4)$$

It follows that, so long as the β_s are regarded as unknown parameters unconnected with Δ, the least squares estimate of Δ depends only on the differences D_s and is in fact the mean of the differences,

$$\hat{\Delta} = \bar{D}_. = \bar{Y}_{T.} - \bar{Y}_{C.}, \qquad (3.5)$$

with

$$\operatorname{var}(\hat{\Delta}) = \operatorname{var}(D_s)/r = 2\sigma^2/r, \qquad (3.6)$$

where σ^2 is the variance of ϵ. Finally σ^2 is estimated as

$$s^2 = \Sigma(D_s - \bar{D}_.)^2/\{2(r-1)\}, \qquad (3.7)$$

so that

$$\operatorname{evar}(\hat{\Delta}) = 2s^2/r. \qquad (3.8)$$

In line with the discussion in Section 2.2.4 we now show that the properties just established under the linear model and the second moment assumption also follow from the randomization used in allocating treatments to units, under the unit-treatment additivity assumption. This assumption specifies the response on the sth pair to be $(\xi_{1s} + \delta, \xi_{2s} - \delta)$ if the first unit in that pair is randomized to treatment and $(\xi_{1s} - \delta, \xi_{2s} + \delta)$ if it is randomized to control. We

MATCHED PAIRS

then have
$$E_R(\hat{\Delta}) = \Delta, \quad E_R\{\text{evar}(\hat{\Delta})\} = \text{var}_R(\hat{\Delta}). \tag{3.9}$$

To prove the second result we note that both sides of the equation do not depend on Δ and are quadratic functions of the ξ_{js}. They are invariant under permutations of the numbering of the pairs $1, \ldots, r$, and under permutations of the two units in any pair. Both sides are zero if $\xi_{1s} = \xi_{2s}, s = 1, \ldots, r$. It follows that both sides of the equation are constant multiples of

$$\Sigma(\xi_{1s} - \xi_{2s})^2 \tag{3.10}$$

and consistency with the least squares analysis requires that the constants of proportionality are equal. In fact, for example,

$$E_R(s^2) = \Sigma(\xi_{1s} - \xi_{2s})^2/(2r). \tag{3.11}$$

Although not necessary for the discussion of the matched pair design, it is helpful for later discussion to set out the relation with analysis of variance. In terms of the original responses Y the estimation of μ, β_s is orthogonal to the estimation of Δ and the analysis of variance arises from the following decompositions.

First there is a representation of the originating random observations in the form

$$\begin{aligned}
Y_{Ts} &= \bar{Y}_{..} + (\bar{Y}_{T.} - \bar{Y}_{..}) + (\bar{Y}_{.s} - \bar{Y}_{..}) \\
&\quad + (\bar{Y}_{Ts} - \bar{Y}_{T.} - \bar{Y}_{.s} + \bar{Y}_{..}), \tag{3.12} \\
Y_{Cs} &= \bar{Y}_{..} + (\bar{Y}_{C.} - \bar{Y}_{..}) + (\bar{Y}_{.s} - \bar{Y}_{..}) \\
&\quad + (\bar{Y}_{Cs} - \bar{Y}_{C.} - \bar{Y}_{.s} + \bar{Y}_{..}). \tag{3.13}
\end{aligned}$$

Regarded as a decomposition of the full vector of observations, this has orthogonal components.

Secondly because of that orthogonality the squared norms of the components add to give

$$\Sigma Y_{js}^2 = \Sigma \bar{Y}_{..}^2 + \Sigma(\bar{Y}_{j.} - \bar{Y}_{..})^2 + \Sigma(\bar{Y}_{.s} - \bar{Y}_{..})^2 + \Sigma(Y_{js} - \bar{Y}_{j.} - \bar{Y}_{.s} + \bar{Y}_{..})^2 : \tag{3.14}$$

note that Σ represents a sum over all observations so that, for example, $\Sigma \bar{Y}_{..}^2 = 2r\bar{Y}_{..}^2$. In this particular case the sums of squares can be expressed in simpler forms. For example the last term is $\Sigma(D_s - \bar{D}_.)^2/2$. The squared norms on the right-hand side are conventionally called respectively sums of squares for general mean, for treatments, for pairs and for residual or error.

Thirdly the dimensions of the spaces spanned by the component vectors, as the vector of observations lies in the full space of dimension $2r$, also are additive:

$$2r = 1 + 1 + (r-1) + (r-1). \tag{3.15}$$

These are conventionally called *degrees of freedom* and *mean squares* are defined for each term as the sum of squares divided by the degrees of freedom. Finally, under the physical linear model (3.3) the residual mean square has expectation σ^2.

3.3.2 A modified matched pair design

In some matched pairs experiments we might wish to include some pairs of units both of which receive the same treatment. Cost considerations might sometimes suggest this as a preferable design, although in that case redefinition of an experimental unit as a pair of original units would be called for and the use of a mixture of designs would not be entirely natural. If, however, there is some suspicion that the two units in a pair do not react independently, i.e. there is doubt about one of the fundamental assumptions of unit-treatment additivity, then a mixture of matched pairs and pairs both treated the same might be appropriate.

Illustration. An opthalmological use of matched pairs might involve using left and right eyes as distinct units, assigning different treatments to the two eyes. This would not be a good design unless there were firm *a priori* grounds for considering that the treatment applied to one eye had negligible influence on the response in the other eye. Nevertheless as a check it might be decided for some patients to assign the same treatment to both eyes, in effect to see whether the treatment difference is the same in both environments. Such checks are, however, often of low sensitivity.

Consider a design in which the r matched pairs are augmented by m pairs in which both units receive the same treatment, m_T pairs receiving T and m_C receiving C, with $m_T + m_C = m$. So long as the parameters β_s in the matched pairs model describing inter-pair differences are arbitrary the additional observations give no information about the treatment effect. In particular a comparison of the means of the m_T and the m_C complete pairs estimates Δ plus a contrast of totally unknown β's.

Suppose, however, that the pairs are randomized between com-

MATCHED PAIRS   47

plete and incomplete assignments. Then under randomization analysis the β's can be regarded in effect as random variables. In terms of a corresponding physical model we write for each observation

$$Y_{js} = \mu \pm \delta + \beta_s + \epsilon_{js}, \qquad (3.16)$$

where the sign of δ depends on the treatment involved, the β_s are now zero mean random variables of variance σ_B^2 and the ϵ_{js} are, as before, zero mean random variables of variance now denoted by σ_W^2. All random variables are mutually uncorrelated or, in the normal theory version, independently normally distributed.

It is again convenient to replace the individual observations by sums and differences. An outline of the analysis is as follows. Let Δ_{MP} and Δ_{UM} denote treatment effects in the matched pairs and the unmatched data respectively. These are estimated by the previous estimate, now denoted by $\bar{Y}_{\mathrm{MPT}} - \bar{Y}_{\mathrm{MPC}}$, with variance $2\sigma_W^2/r$ and by $\bar{Y}_{\mathrm{UMT}} - \bar{Y}_{\mathrm{UMC}}$ with variance

$$(\sigma_B^2 + \sigma_W^2/2)(1/m_T + 1/m_C). \qquad (3.17)$$

If, as might quite often be the case, σ_B^2 is large compared with σ_W^2, the between block comparison may be of such low precision as to be virtually useless.

If the variance components are known we can thus test the hypothesis that the treatment effect is, as anticipated *a priori*, the same in the two parts of the experiment and subject to homogeneity find a weighted mean as an estimate of the common Δ. Estimation of the two variance components is based on the sum of squares within pairs adjusting for treatment differences in the matched pair portion and on the sum of squares between pair totals adjusting for treatment differences in the unmatched pair portion.

Under normal theory assumptions a preferable analysis for a common Δ is summarized in Exercise 3.3. There are five sufficient statistics, two sums of squares and three means, and four unknown parameters. The log likelihood of these statistics can be found and a profile log likelihood for Δ calculated.

The procedure of combining information from within and between pair comparisons can be regarded as the simplest special case of the recovery of between-block information. More general cases are discussed in Section 4.2.

3.4 Randomized block design

3.4.1 Model and analysis

Suppose now that we have more than two treatments and that they are regarded as unstructured and on an equal footing and therefore to be equally replicated. The discussion extends in a fairly direct way when some treatments receive additional replication. With v treatments, or varieties in the plant breeding context, we aim to produce blocks of v units. As with matched pairs we try, subject to administrative constraints on the experiment, to arrange that in the absence of treatment effects, very similar responses are to be anticipated on the units within any one block. We allocate treatments independently from block to block and at random within each block, subject to the constraint that each treatment occurs once in each block.

Illustration. Typical ways of forming blocks include compact arrangements of plots in a field chosen in the light of any knowledge about fertility gradients, batches of material that can be produced in one day or production period, and animals grouped on the basis of gender and initial body weight.

Let Y_{js} denote the observation on treatment T_j in block s. Note that because of the randomization this observation may be on any one of the units in block s in their original listing. In accordance with the general principle that constraints on the randomization are represented by parameters in the associated linear model, we represent Y_{js} in the form

$$Y_{js} = \mu + \tau_j + \beta_s + \epsilon_{js}, \qquad (3.18)$$

where $j = 1, \ldots, v$; $s = 1, \ldots, r$ and ϵ_{js} are zero mean random variables satisfying the second moment or normal theory assumptions. The least squares estimates of the parameters are determined by the row and column means and in particular under the summation constraints $\Sigma \tau_j = 0$, $\Sigma \beta_s = 0$, we have $\hat{\tau}_j = \bar{Y}_{j.} - \bar{Y}_{..}$ and $\hat{\beta}_s = \bar{Y}_{.s} - \bar{Y}_{..}$. The contrast $L_\tau = \Sigma l_j \tau_j$ is estimated by $\hat{L}_\tau = \Sigma l_j \bar{Y}_{j.}$.

The decomposition of the observations, the sums of squares and the degrees of freedom are as follows:

1. For the observations we write

$$\begin{aligned} Y_{js} &= \bar{Y}_{..} + (\bar{Y}_{j.} - \bar{Y}_{..}) + (\bar{Y}_{.s} - \bar{Y}_{..}) \\ &\quad + (Y_{js} - \bar{Y}_{j.} - \bar{Y}_{.s} + \bar{Y}_{..}), \end{aligned} \qquad (3.19)$$

a decomposition into orthogonal components.

2. For the sums of squares we therefore have
$$\Sigma Y_{js}^2 = \Sigma \bar{Y}_{..}^2 + \Sigma(\bar{Y}_{j.} - \bar{Y}_{..})^2 + \Sigma(\bar{Y}_{.s} - \bar{Y}_{..})^2 \\ + \Sigma(Y_{js} - \bar{Y}_{j.} - \bar{Y}_{.s} + \bar{Y}_{..})^2, \quad (3.20)$$
where the summation is always over both suffices.

3. For the degrees of freedom we have
$$rv = 1 + (v-1) + (r-1) + (r-1)(v-1). \quad (3.21)$$

The residual mean square provides an unbiased estimate of the variance. Let
$$s^2 = \Sigma(Y_{js} - \bar{Y}_{j.} - \bar{Y}_{.s} + \bar{Y}_{..})^2 / \{(r-1)(v-1)\}. \quad (3.22)$$
We now indicate how to establish the result $E(s^2) = \sigma^2$ under the second moment assumptions. In the linear model the residual sum of squares depends only on $\{\epsilon_{js}\}$, and not on the fixed parameters μ, $\{\tau_j\}$ and $\{\beta_s\}$. Thus for the purpose of computing the expected value of (3.22) we can set these parameters to zero. All sums of squares in (3.20) other than the residual have simple expectations: for example
$$E\{\Sigma_{j,s}(\bar{Y}_{j.} - \bar{Y}_{..})^2\} = rE\{\Sigma_j(\bar{\epsilon}_{j.} - \bar{\epsilon}_{..})^2\} \quad (3.23)$$
$$= r(v-1)\text{var}(\bar{\epsilon}_{j.}) = (v-1)\sigma^2. \quad (3.24)$$
Similarly $E\{\Sigma_{j,s}(\bar{Y}_{.s} - \bar{Y}_{..})^2\} = (r-1)\sigma^2$, $E(\Sigma_{j,s}\bar{Y}_{..}^2) = \sigma^2$, and that for the residual sum of squares follows by subtraction. Thus the unbiased estimate of the variance of $\hat{L}_\tau$ is
$$\text{evar}(\hat{L}_\tau) = \Sigma_j l_j^2 s^2 / r. \quad (3.25)$$

The partition of the sums of squares given by (3.20) is often set out in an analysis of variance table, as for example Table 3.2 below. This table has one line for each component of the sum of squares, with the usual convention that the sums of squares due to the overall mean, $n\bar{Y}_{..}^2$, is not displayed, and the total sum of squares is thus a corrected total $\Sigma(Y_{js} - \bar{Y}_{..})^2$.

The simple decomposition of the data vector and sum of squares depend crucially on the balance of the design. If, for example, some treatments were missing in some blocks not merely would the orthogonality of the component vectors be lost but the contrasts of treatment means would not be independent of differences between blocks and *vice versa*. To extend the discussion to such cases more elaborate methods based on a least squares analysis are needed. It becomes crucial to distinguish, for example, between the sum of squares for treatments ignoring blocks and the sum of squares for treatments adjusting for blocks, the latter measuring the effect of introducing treatment effects after first allowing for block differences.

The randomization model for the randomized block design uses the assumption of unit-treatment additivity, as in the matched pairs design. We label the units

$$U_{11},\ldots,U_{v1}; \quad U_{12},\ldots,U_{v2}; \quad \ldots; \quad U_{1r},\ldots,U_{vr}. \qquad (3.26)$$

The response on the unit in the sth block that is randomized to treatment T_j is

$$\xi_{T_j s} + \tau_j \qquad (3.27)$$

where $\xi_{T_j s}$ is the response of that unit in block s in the absence of treatment.

Under randomization theory properties such as

$$E_R\{\text{evar}(\hat{L}_\tau)\} = \text{var}_R(\hat{L}_\tau) \qquad (3.28)$$

are established by first showing that both sides are multiples of

$$\Sigma(\xi_{js} - \bar{\xi}_{.s})^2. \qquad (3.29)$$

3.4.2 Example

This example is taken from Cochran and Cox (1958, Chapter 3), and is based on an agricultural field trial. In such trials blocks are naturally formed from large sections of field, sometimes roughly square; the shape of individual plots and their arrangement into plots is usually settled by a mixture of technological convenience, for example ease of harvesting, and special knowledge of the particular area.

This experiment tested the effects of five levels of application of potash on the strength of cotton fibres. A single sample of cotton was taken from each plot, and four measurements of strength were made on each sample. The data in Table 3.1 are the means of these four measurements.

The marginal means are given in Table 3.1, and seem to indicate decreasing strength with increasing amount of potash, with perhaps some curvature in the response, since the mean strength

RANDOMIZED BLOCK DESIGN

Table 3.1 *Strength index of cotton, from Cochran and Cox (1958), with marginal means.*

		\multicolumn{5}{c}{Pounds of potash per acre}					
		36	54	72	108	144	Mean
	I	7.62	8.14	7.76	7.17	7.46	7.63
Block	II	8.00	8.15	7.73	7.57	7.68	7.83
	III	7.93	7.87	7.74	7.80	7.21	7.71
Mean		7.85	8.05	7.74	7.51	7.45	7.72

Table 3.2 *Analysis of variance for strength index of cotton.*

Source	Sums of squares	Degrees of freedom	Mean square
Treatment	0.7324	4	0.1831
Blocks	0.0971	2	0.0486
Residual	0.3495	8	0.0437

at 36 pounds is less than that at 54 pounds, where the maximum is reached.

The analysis of variance outlined in Section 3.4.1 is given in Table 3.2. The main use of the analysis of variance table is to provide an estimate of the standard error for assessing the precision of contrasts of the treatment means. The mean square residual is an unbiased estimate of the variance of an individual observation, so the standard error for example for comparing two treatment means is $\sqrt{(2 \times 0.0437/3)} = 0.17$, which suggests that the observed decrease in strength over the levels of potash used is a real effect, but the observed initial increase is not.

It is possible to construct more formal tests for the shape of the response, by partitioning the sums of squares for treatments, and this is considered further in Section 3.5 below.

The S-PLUS code for carrying out the analysis of variance in this and the following examples is given in Appendix C. As with many other statistical packages, the emphasis in the basic commands is on the analysis of variance table and the associated F-tests, which in nearly all cases are not the most useful summary information.

3.4.3 Efficiency of blocking

As noted above the differences between blocks are regarded as of no intrinsic interest, so long as no relevant baseline information is available about them. Sometimes, however, it may be useful to ask how much gain in efficiency there has been as compared with complete randomization. The randomization model provides a means of assessing how effective the blocking has been in improving precision. In terms of randomization theory the variance of the difference between two treatment means in a completely randomized experiment is determined by

$$\frac{2}{r}\Sigma(\xi_{js} - \bar{\xi}_{..})^2/(vr - 1), \qquad (3.30)$$

whereas in the randomized block experiment it is

$$\frac{2}{r}\Sigma(\xi_{js} - \bar{\xi}_{.s})^2/\{r(v - 1)\}. \qquad (3.31)$$

Also in the randomization model the mean square between blocks is constant with value

$$v\Sigma(\bar{\xi}_{.s} - \bar{\xi}_{..})^2/(r - 1). \qquad (3.32)$$

As a result the relative efficiency for comparing two treatment means in the two designs is estimated by

$$\frac{2}{r}\frac{\text{SS}_B + r(v - 1)\text{MS}_R}{(vr - 1)\text{MS}_R}. \qquad (3.33)$$

Here SS_B and MS_R are respectively the sum of squares for blocks and the residual mean square in the original randomized block analysis.

To produce from the original analysis of variance table for the randomized block design an estimate of the effective residual variance for the completely randomized design we may therefore produce a new formal analysis of variance table as follows. Replace the treatment mean square by the residual mean square, add the sums of squares for modified treatments, blocks and residual and divide by the degrees of freedom, namely $vr - 1$. The ratio of the two residual mean squares, the one in the analysis of the randomized

PARTITIONING SUMS OF SQUARES

block experiment to the notional one just reconstructed, measures the reduction in effective variance induced by blocking.

There is a further aspect, however; if confidence limits for Δ are found from normal theory using the Student t distribution, the degrees of freedom are $(v-1)(r-1)$ and $v(r-1)$ respectively in the randomized block and completely randomized designs, showing some advantage to the latter if the error variances remain the same. Except in very small experiments, however, this aspect is relatively minor.

3.5 Partitioning sums of squares

3.5.1 General remarks

We have in this chapter emphasized that the objective of the analysis is the estimation of comparisons between the treatments. In the context of analysis of variance the sum of squares for treatments is a summary measure of the variation between treatments and could be the basis of a test of the overall null hypothesis that all treatments have identical effect, i.e. that the response obtained on any unit is unaffected by the particular treatment assigned to it. Such a null hypothesis is, however, very rarely of concern and therefore the sum of squares for treatments is of importance primarily in connection with the computation of the residual sum of squares, the basis for estimating the error variance.

It is, however, important to note that the treatment sum of squares can be decomposed into components corresponding to comparisons of the individual effects and this we now develop.

3.5.2 Contrasts

Recall from Section 2.2.6 that if the treatment parameters are denoted by $\tau_1, \ldots, \tau_v$ a linear combination $L_\tau = \Sigma l_j \tau_j$ is called a treatment contrast if $\Sigma l_j = 0$. The contrast L_τ is estimated in the randomized block design by

$$\hat{L}_\tau = \Sigma_j l_j \bar{Y}_{j.}, \qquad (3.34)$$

where $\bar{Y}_{j.}$ is the mean response on the jth treatment, averaged over blocks. Equivalently we can write

$$\hat{L}_\tau = \Sigma_{j,s} l_j Y_{js}/r, \qquad (3.35)$$

where the sum is over individual observations and r is the number of replications of each treatment.

Under the linear model (3.18) and the second moment assumption,

$$E(\hat{L}_\tau) = L_\tau, \quad \mathrm{var}(\hat{L}_\tau) = \sigma^2 \Sigma_j l_j^2 / r. \tag{3.36}$$

We now define the sum of squares with one degree of freedom associated with L_τ to be

$$\mathrm{SS}_L = r\hat{L}_\tau^2 / \Sigma l_j^2. \tag{3.37}$$

This definition is in some ways most easily recalled by noting that $\hat{L}_\tau$ is a linear combination of responses, and hence SS_L is the squared length of the orthogonal projection of the observation vector onto the vector whose components are determined by l.

The following properties are derived directly from the definitions:
1. $E(\hat{L}_\tau) = L_\tau$ and is zero if and only if the population contrast is zero.
2. $E(\mathrm{SS}_L) = \sigma^2 + rL_\tau^2/\Sigma l_j^2$.
3. Under the normal theory assumption SS_L is proportional to a noncentral chi-squared random variable with one degree of freedom reducing to the central chi-squared form if and only if $L_\tau = 0$.
4. The square of the Student t statistic for testing the null hypothesis $L_\tau = 0$ is the analysis of variance F statistic for comparing SS_L with the residual mean square.

In applications the Student t form is to be preferred to its square, partly because it preserves the information in the sign and more importantly because it leads to the determination of confidence limits.

3.5.3 Mutually orthogonal contrasts

Several contrasts $L_\tau^{(1)}, L_\tau^{(2)}, \ldots$ are called mutually orthogonal if for all $p \neq q$

$$\Sigma l_j^{(p)} l_j^{(q)} = 0. \tag{3.38}$$

Note that under the normal theory assumption the estimates of orthogonal contrasts are independent. The corresponding Student t statistics are not quite independent because of the use of a common estimate of σ^2, although this is a minor effect unless the residual degrees of freedom are very small.

PARTITIONING SUMS OF SQUARES

Now suppose that there is a complete set of $v-1$ mutually orthogonal contrasts. Then by forming an orthogonal transformation of $\bar{Y}_{1.},\ldots,\bar{Y}_{v.}$ from $(1/\sqrt{v},\ldots,1/\sqrt{v})$ and the normalized contrast vectors, it follows that

$$r\Sigma_{js}(\bar{Y}_{j.} - \bar{Y}_{..})^2 = \mathrm{SS}_{L_\tau^{(1)}} + \ldots + \mathrm{SS}_{L_\tau^{(v)}}, \qquad (3.39)$$

that is the treatment sum of squares has been decomposed into single degrees of freedom.

Further if there is a smaller set of $v_1 < v-1$ mutually orthogonal contrasts, then the treatment sum of squares can be decomposed into

Selected individual contrasts	v_1
Remainder	$v - 1 - v_1$
Total for treatments	$v - 1$

In this analysis comparison of the mean square for the remainder term with the residual mean square tests the hypothesis that all treatment effects are accounted for within the space of the v_1 identified contrasts. Thus with six treatments and the single degree of freedom contrasts identified by

$$L_\tau^{(1)} = (\tau_1 + \tau_2)/2 - \tau_3, \qquad (3.40)$$
$$L_\tau^{(2)} = (\tau_1 + \tau_2 + \tau_3)/3 - (\tau_4 + \tau_5 + \tau_6)/3, \qquad (3.41)$$

we have the partition

$L_\tau^{(1)}$	1
$L_\tau^{(2)}$	1
Remainder	3
Total for treatments	5

The remainder term could be divided further, perhaps most naturally initially into a contrast of τ_1 with τ_2 and a comparison with two degrees of freedom among the last three treatments.

The orthogonality of the contrasts is required for the simple decomposition of the sum of squares. Subject-matter relevance of the comparisons of course overrides mathematical simplicity and it may be unavoidable to look at nonorthogonal comparisons.

We have in this section used notation appropriate to partitioning the treatment sums of squares in a randomized block design, but the same ideas apply directly to more general settings, with $\bar{Y}_{j.}$ above replaced by the average of all observations on the jth treatment, and r replaced by the number of replications of each treatment. When in Chapter 5 we consider more complex treatments defined by factors exactly the same analysis can be applied to interactions.

3.5.4 Equally spaced treatment levels

A particularly important special case arises when treatments are defined by levels of a quantitative variable, often indeed by equally spaced values of that variable. For example a dose might be set at four levels defined by log dose $= 0, 1, 2, 3$ on some suitable scale, or a temperature might have three levels defined by temperatures of 30, 40, 50 degrees Celsius, and so on.

We now discuss the partitioning of the sums of squares for such a quantitative treatment in orthogonal components, corresponding to regression on that variable. It is usual, and sensible, with quantitative factors at equally spaced levels, to use contrasts representing linear, quadratic, cubic, ... dependence of the response on the underlying variable determining the factor levels. Tables of these contrasts are widely available and are easily constructed from first principles via orthogonal polynomials, i.e. via Gram-Schmidt orthogonalization of $\{1, x, x^2, \ldots\}$. For a factor with three equally spaced levels, the linear and quadratic contrasts are

$$\begin{array}{ccc} -1 & 0 & 1 \\ 1 & -2 & 1 \end{array}$$

and for one with four equally spaced levels, the linear, quadratic and cubic contrasts are

$$\begin{array}{cccc} -3 & -1 & 1 & 3 \\ 1 & -1 & -1 & 1 \\ -1 & 3 & 3 & 1 \end{array}$$

The sums of squares associated with these can be compared with the appropriate residual sum of squares. In this way some notion of the shape of the dependence of the response on the variable defining the factor can be obtained.

3.5.5 Example 3.4 continued

In this example the treatments were defined by increasing levels of potash, in pounds per acre. The levels used were 36, 54, 72, 108 and 144. Of interest is the shape of the dependence of strength on level of potash; there is some indication in Table 3.1 of a levelling off or decrease of response at the highest level of potash.

These levels are not equally spaced, so the orthogonal polynomials of the previous subsection are not exactly correct for extracting linear, quadratic, and other components. The most accurate way of partitioning the sums of squares for treatments is to use regression methods or equivalently to construct the appropriate orthogonal polynomials from first principles. We will illustrate here the use of the usual contrasts, as the results are much the same.

The coefficients for the linear contrast with five treatment levels are $(-2, -1, 0, 1, 2)$, and the sum of squares associated with this contrast is $SS_{lin} = 3(-1.34)^2/10 = 0.5387$. The nonlinear contribution to the treatment sum of squares is thus just 0.1938 on three degrees of freedom, which indicates that the suggestion of nonlinearity in the response is not significant. The quadratic component, defined by the contrast $(2, -1, -2, -1, 2)$ has an associated sum of squares of 0.0440.

If we use the contrast exactly appropriate for a linear regression, which has entries proportional to

$$(-2, -1.23, -0.46, 1.08, 2.61),$$

we obtain the same conclusion.

With more extensive similar data, or with various sets of similar data, it would probably be best to fit a nonlinear model consistent with general subject-matter knowledge, for example an exponential model rising to an asymptote. Fitting such a model across various sets of data should be helpful for the comparison and synthesis of different studies.

3.6 Retrospective adjustment for improving precision

In Section 3.1 we reviewed various ways of improving precision and in Sections 3.2 and 3.3 developed the theme of comparing like with like via blocking the experimental units into relatively homogeneous sets, using baseline information. We now turn to a second use of baseline information. Suppose that on each experimental unit there is a vector z of variables, either quantitative or indicators of qualitative groupings and that this information has either not been used in forming blocks or at least has been only partly used.

There are three rather different situations. The importance of z may have been realized only retrospectively, for example by an investigator different from the one involved in design. It may have been more important to block on features other than z; this is especially relevant when a large number of baseline features is available. Thirdly, any use of z to form blocks is qualitative and it may be that quantitative use of z instead of, or as well as, its use to form blocks may add sensitivity.

Illustrations. In many clinical trials there will be a large number of baseline features available at the start of a trial and the practicalities of randomization may restrict blocking to one or two key features such as gender and age or gender and initial severity. In an animal experiment comparing diets, blocks could be formed from animals of the same gender and roughly the same initial body weight but, especially in small experiments, appreciable variation in initial body weight might remain within blocks.

Values of z can be used to test aspects of unit-treatment additivity, in effect via tests of parallelism, but here we concentrate on precision improvement. The formal statistical procedures of introducing regression on z into a model have appeared in slightly different guise in Section 2.3 as techniques for retrospective bias removal and will not be repeated. In fact what from a design perspective is random error can become bias at the stage of analysis, when conditioning on relevant baseline features is appropriate. It is therefore not surprising that the same statistical technique reappears.

Illustration. A group of animals with roughly equal numbers of males and females is randomized between two treatments T and C regardless of gender. It is then realized that there are substantially more males than females in T. From an initial design perspective this is a random fluctuation: it would not persist in a similar large study. On the other hand once the imbalance is observed, unless it can be dismissed as irrelevant or unimportant it is a potential source of bias and is to be removed by rerandomizing or, if it is too late for that, by appropriate analysis. This aspect is connected with

RETROSPECTIVE ADJUSTMENT FOR IMPROVING PRECISION

some difficult conceptual issues about randomization; see Section 2.4.

This discussion raises at least two theoretical issues. The first concerns the possible gains from using a single quantitative baseline variable both as a basis for blocking and after that also as a basis for an adjustment. It can be shown that only when the correlation between baseline feature and response is very high is this double use of it likely to lead to a substantial improvement in final precision.

Suppose now that there are baseline features that cannot be reasonably controlled by blocking and that they are controlled by a regression adjustment. Is there any penalty associated with adjusting unnecessarily?

To study this consider first an experiment to compare two treatments, with r replicates of each. After adjustment for the $q \times 1$ vector of baseline variables, z, the variance of the estimated difference between the treatments is

$$\operatorname{var}(\hat{\tau}_T - \hat{\tau}_C) = \sigma^2 \{2/r + (\bar{z}_{T.} - \bar{z}_{C.})^T R_{zz}^{-1} (\bar{z}_{T.} - \bar{z}_{C.})\}, \quad (3.42)$$

where σ^2 is the variance per observation residual to regression on z and to any blocking system used, $\bar{z}_{T.}, \bar{z}_{C.}$ are the treatment mean vectors and R_{zz} is the matrix of sums of squares and cross-products of z within treatments again eliminating any block effects.

Now if treatment assignment is randomized

$$E_R(R_{zz}/d_w) = \Omega_{zz}, \quad (3.43)$$

where d_w is the degrees of freedom of the residual sum of squares in the analysis of variance table, and Ω_{zz} is a finite population covariance matrix of the unit constants within blocks. With $v = 2$ we have

$$E_R(\bar{z}_T - \bar{z}_C) = 0, \quad E_R\{(\bar{z}_T - \bar{z}_C)(\bar{z}_T - \bar{z}_C)^T\} = 2\Omega_{zz}/r. \quad (3.44)$$

Now

$$\frac{1}{2} r (\bar{z}_T - \bar{z}_C)^T \Omega_{zz}^{-1} (\bar{z}_T - \bar{z}_C) = \frac{1}{2} r \|\bar{z}_T - \bar{z}_C\|_{\Omega_{zz}}^2, \quad (3.45)$$

say, has expectation q and approximately a chi-squared distribution with q degrees of freedom.

That is, approximately

$$\operatorname{var}(\hat{\tau}_T - \hat{\tau}_C) = \frac{2\sigma^2}{r}(1 + W_q/d_w), \quad (3.46)$$

where W_q denotes a random variable depending on the outcome of the randomization and having approximately a chi-squared distribution with q degrees of freedom.

More generally if there are v treatments each replicated r times

$$\text{ave}_{j\neq l}\text{var}(\hat\tau_j - \hat\tau_l) = \sigma^2[2/r + 2/\{r(v-1)\}\text{tr}(B_{zz}R_{zz}^{-1})], \quad (3.47)$$

where B_{zz} is the matrix of sums of squares and products between treatments and $\text{tr}(A)$ denotes the trace of the matrix A, i.e. the sum of the diagonal elements.

The simplest interpretation of this is obtained by replacing W_q by its expectation, and by supposing that the number of units n is large compared with the number of treatments and blocks, so that $d_w \sim n$. Then the variance of an estimated treatment difference is approximately

$$\frac{2\sigma^2}{r}(1 + \frac{q}{n}). \quad (3.48)$$

The inflation factor relative to the randomized block design is approximately $n/(n-q)$ leading to the conclusion that every unnecessary parameter fitted, i.e. adjustment made without reduction in the effective error variance per unit, σ^2, is equivalent to the loss of one experimental unit.

This conclusion is in some ways oversimplified, however, not only because of the various approximations in its derivation. First, in a situation such as a clinical trial with a potentially large value of q, adjustments would be made selectively in a way depending on the apparent reduction of error variance achieved. This makes assessment more difficult but the inflation would probably be rather more than that based on q_0, the dimension of the z actually used, this being potentially much less than q, the number of baseline features available.

The second point is that the variance inflation, which arises because of the nonorthogonality of treatments and regression analyses in the least squares formulation, is a random variable depending on the degree of imbalance in the configuration actually used. Now if this imbalance can be controlled by design, for example by rerandomizing until the value of W_q is appreciably smaller than its expectation, the consequences for variance inflation are reduced and possibly but not necessarily the need to adjust obviated. If, however, such control at the design stage is not possible, the average inflation may be a poor guide. It is unlikely though that the

SPECIAL MODELS OF ERROR VARIATION   61

inflation will be more for small ϵ than

$$(1 + W_{q,\epsilon}/n), \tag{3.49}$$

where $W_{q,\epsilon}$ is the upper ϵ point of the randomization distribution of W_q, approximately a chi-squared distribution with q degrees of freedom.

For example, with $\epsilon = 0.01$ and $q = 10$ it will be unlikely that there is more than a 10 per cent inflation if $n > 230$ as compared with $n > 100$ suggested by the analysis based on properties averaged over the randomization distribution. Note that when the unadjusted and adjusted effects differ immaterially simplicity of presentation may favour the former.

A final point concerns the possible justification of the adjusted analysis based on randomization and the assumption of unit treatment additivity. Such a justification is usually only approximate but can be based on an approximate conditional distribution regarding, in the simplest case of just two treatments, $\bar{z}_T - \bar{z}_C$ as fixed.

3.7 Special models of error variation

In this chapter we have emphasized methods of error control by blocking which, combined with randomization, aim to increase the precision of estimated treatment contrasts without strong special assumptions about error structure. That is, while the effectiveness of the methods in improving precision depends on the way in which the blocks are formed, and hence on prior knowledge, the validity of the designs and the associated standard errors does not do so.

Sometimes, however, especially in relatively small experiments in which the experimental units are ordered in time or systematically arrayed in space a special stochastic model may reasonably be used to represent the error variation. Then there is the possibility of using a design that exploits that model structure. However, usually the associated method of analysis based on that model will not have a randomization justification and we will have to rely more strongly on the assumed model than for the designs discussed in this chapter.

When the experimental units are arranged in time the two main types of variation are a trend in time supplemented by totally random variation and a stationary time series representation. The latter is most simply formulated via a low order autoregression.

For spatial problems there are similar rather more complex representations. Because the methods of design and analysis associated with these models are more specialized we defer their discussion to Chapter 8.

3.8 Bibliographic notes

The central notions of blocking and of adjustment for baseline variables are part of the pioneering contributions of Fisher (1935), although the qualitative ideas especially of the former have a long history. The relation between the adjustment process and randomization theory was discussed by Cox (1982). See also the Bibliographic notes to Chapter 2. For the relative advantages of blocking and adjustment via a baseline variable, see Cox (1957).

The example in Section 3.4 is from Cochran and Cox (1958, Chapter 3), and the partitioning of the treatment sum of squares follows closely their discussion. The analysis of matched pairs and randomized blocks from the linear model is given in most books on design and analysis; see, for example, Montgomery (1997, Chapters 2 and 5) and Dean and Voss (1999, Chapter 10). The randomization analysis is given in detail in Hinkelmann and Kempthorne (1994, Chapter 9), as is the estimation of the efficiency of the randomized block design, following an argument attributed to Yates (1937).

3.9 Further results and exercises

1. Under what circumstances would it be reasonable to have a randomized block experiment in which each treatment occurred more than once, say, for example, twice, in each block, i.e. in which the number of units per block is twice the number of treatments? Set out the analysis of variance table for such a design and discuss what information is available that cannot be examined in a standard randomized block design.

2. Suppose in a matched pair design the responses are binary. Construct the randomization test for the null hypothesis of no treatment difference. Compare this with the test based on that for the binomial model, where Δ is the log odds-ratio. Carry out a similar comparison for responses which are counts of numbers of occurrences of point events modelled by the Poisson distribution.

FURTHER RESULTS AND EXERCISES

3. Consider the likelihood analysis under the normal theory assumptions of the modified matched pair design of Section 3.3.2. There are r matched pairs, m_T pairs in which both units receive T and m_C pairs in which both units receive C; we assume a common treatment difference applies throughout. We transform the original pairs of responses to sums and differences as in Section 3.3.1.

(a) Show that r of the differences have mean Δ, and that $m_T + m_C$ of them have mean zero, all differences being independently normally distributed with variance τ_D, say.

(b) Show that independently of the differences the sums are independently normally distributed with variance τ_S, say, with r having mean ν, say, m_T having mean $\nu + \delta$ and m_C having mean $\nu - \delta$, where $\Delta = 2\delta$.

(c) Hence show that minimal sufficient statistics are (i) the least squares estimate of ν from the sums; (ii) the least squares estimate $\hat{\Delta}_S$ of Δ from the unmatched pairs, i.e. the difference of the means of m_T and m_C pairs; (iii) the estimate $\hat{\Delta}_D$ from the matched pairs; (iv) a mean square MS_D with $d_D = r - 1 + m_T + m_C$ degrees of freedom estimating τ_D and (v) a mean square MS_S with $d_S = r - 2 + m_T + m_C$ degrees of freedom estimating τ_S. This shows that the system is a $(5, 4)$ curved exponential family.

(d) Without developing a formal connection with randomization theory note that complete randomization of pairs to the three groups would give some justification to the strong homogeneity assumptions involved in the above. How would such homogeneity be examined from the data?

(e) Show that a log likelihood function obtained by ignoring (i) and using the known densities of the four remaining statistics is

$$-\frac{1}{2}\log \tau_S - \tilde{m}(\hat{\Delta}_S - \Delta)^2/(2\tau_S)$$
$$-\frac{1}{2}\log \tau_D - r(\hat{\Delta}_D - \Delta)^2/(2\tau_D)$$
$$-\frac{1}{2}d_D \log \tau_D - \frac{1}{2}d_D \text{MS}_D/\tau_D$$
$$-\frac{1}{2}d_S \log \tau_S - \frac{1}{2}d_S \text{MS}_S/\tau_S,$$

where $1/\tilde{m} = 1/m_D + 1/m_S$.

(f) Hence show, possibly via some simulated data, that only in quite small samples will the profile likelihood for Δ differ appreciably from that corresponding to a weighted combination of the two estimates of Δ replacing the variances and theoretically optimal weights by sample estimates and calculating confidence limits via the Student t distribution with effective degrees of freedom

$$\tilde{d} = (r\text{MS}_S + \tilde{m}\text{MS}_D)^2 (r^2 \text{MS}_S^2/d_D + \tilde{m}^2 \text{MS}_D^2/d_S)^{-1}.$$

For somewhat related calculations, see Cox (1984b).

4. Suppose that n experimental units are arranged in sequence in time and that there is prior evidence that the errors are likely to be independent and identically distributed initially with mean zero except that at some as yet unknown point there is likely to be a shift in mean error. What design would be appropriate for the comparison of v treatments? After the experiment is completed and the responses obtained it is found that the discontinuity has indeed occurred. Under the usual linear assumptions what analysis would be suitable if

(a) the position of the discontinuity can be determined without error from supplementary information

(b) the position of the discontinuity is regarded as an unknown parameter.

CHAPTER 4

Specialized blocking techniques

4.1 Latin squares

4.1.1 Main ideas

In some fields of application it is quite common to have two different qualitative criteria for grouping units into blocks, the two criteria being cross-classified. That is, instead of the units being conceptually grouped into sets or blocks, arbitrarily numbered in each block, it may be reasonable to regard the units as arrayed in a two-dimensional way.

Illustrations. Plots in an agricultural trial may be arranged in a square or rectangular array in the field with both rows and columns likely to represent systematic sources of variation. The shapes of individual plots will be determined by technological considerations such as ease of harvesting.

In experimental psychology it is common to expose each subject to a number of different conditions (treatments) in sequence in each experimental session. Thus with v conditions used per subject per session, we may group the subjects into sets of v. For each set of v subjects the experimental units, i.e. subject-period combinations, form a $v \times v$ array, with potentially important sources of variation both between subjects and between periods. In such experiments where the same individual is used as a unit more than once, the assumption that the response on one unit depends only on the treatment applied to that unit may need close examination.

In an industrial process with similar machines in parallel it may be sensible to regard machines and periods as the defining features of a two-way classification of the units.

The simplest situation arises with v treatments, and two blocking criteria each with v levels. The experimental units are arranged in one or more $v \times v$ squares. Then the principles of comparing like with like and of randomization suggest using a design with each treatment once in each row and once in each column and choosing a design at random subject to those two constraints. Such a design is called a $v \times v$ *Latin square*.

An example of a 4×4 Latin square after randomization is

$$\begin{array}{cccc} T_4 & T_2 & T_3 & T_1 \\ T_2 & T_4 & T_1 & T_3 \\ T_1 & T_3 & T_4 & T_2 \\ T_3 & T_1 & T_2 & T_4. \end{array} \qquad (4.1)$$

In an application in experimental psychology the rows might correspond to subjects and the columns to periods within an experimental session. The arrangement ensures that constant differences between subjects or between periods affect all treatments equally and thus do not induce error in estimated treatment contrasts.

Randomization is most simply achieved by starting with a square in some standardized form, for example corresponding to cyclic permutation:

$$\begin{array}{cccc} A & B & C & D \\ B & C & D & A \\ C & D & A & B \\ D & A & B & C \end{array} \qquad (4.2)$$

and then permuting the rows at random, permuting the columns at random, and finally assigning the letters $A, \ldots, D$ to treatments at random. The last step is unnecessary to achieve agreement between the randomization model and the standard linear model, although it probably ensures a better match between normal theory and randomization-based confidence limits; we do not, however, know of specific results on this issue. It would, for the smaller squares at least, be feasible to choose at random from all Latin squares of the given size, yielding an even richer randomization set.

On the basis again that constraints on the design are to be reflected by parameters in the physical linear model, the default model for the analysis of a Latin square is as follows. In row s and column t let the treatment be $T_{j_{st}}$. Then

$$Y_{j_{st}st} = \mu + \tau_j + \beta_s + \gamma_t + \epsilon_{jst}, \qquad (4.3)$$

where on the right-hand side we have abbreviated j_{st} to j and where the assumptions about ϵ_{jst} are as usual either second moment assumptions or normal theory assumptions. Note especially that on the right-hand side of (4.3) the suffices j, s, t do not range

LATIN SQUARES

freely. The least-squares estimate of the contrast $\Sigma l_j \tau_j$ is

$$\Sigma l_j \bar{Y}_{j..}, \tag{4.4}$$

where $\bar{Y}_{j..}$ is the mean of the responses on T_j. Further

$$\operatorname{evar}(\Sigma l_j \bar{Y}_{j..}) = \Sigma l_j^2 s^2 / v, \tag{4.5}$$

where

$$s^2 = \Sigma(Y_{jst} - \bar{Y}_{j..} - \bar{Y}_{.s.} - \bar{Y}_{..t} + 2\bar{Y}_{...})^2 / \{(v-1)(v-2)\}. \tag{4.6}$$

The justification can be seen most simply from the following decompositions:

1. For the observations

$$Y_{jst} = \bar{Y}_{...} + (\bar{Y}_{j..} - \bar{Y}_{...}) + (\bar{Y}_{.s.} - \bar{Y}_{...}) + (\bar{Y}_{..t} - \bar{Y}_{...})$$
$$+ (Y_{jst} - \bar{Y}_{j..} - \bar{Y}_{.s.} - \bar{Y}_{..t} + 2\bar{Y}_{...}). \tag{4.7}$$

2. For the sums of squares

$$\Sigma Y_{jst}^2 = \Sigma \bar{Y}_{...}^2 + \Sigma(\bar{Y}_{j..} - \bar{Y}_{...})^2 + \Sigma(\bar{Y}_{.s.} - \bar{Y}_{...})^2$$
$$+ \Sigma(\bar{Y}_{..t} - \bar{Y}_{...})^2 + \Sigma(Y_{jst} - \bar{Y}_{j..} - \bar{Y}_{.s.} - \bar{Y}_{..t} + 2\bar{Y}_{...})^2. \tag{4.8}$$

3. For the degrees of freedom

$$v^2 = 1 + (v-1) + (v-1) + (v-1) + (v-1)(v-2). \tag{4.9}$$

The second moment properties under the randomization distribution are established as before. From this point of view the assumption underlying the Latin square design, namely that of unit-treatment additivity, is identical with that for completely randomized and randomized block designs. In Chapter 6 we shall see other interpretations of a Latin square as a fractional factorial experiment and there strong additional assumptions will be involved.

Depending on the precision to be achieved it may be necessary to form the design from several Latin squares. Typically they would be independently randomized and the separate squares kept separate in doing the experiment, the simplest illustration of what is called a *resolvable design*. This allows the elimination of row and column effects separately within each square and also permits the analysis of each square separately, as well as an analysis of the full set of squares together. Another advantage is that if there is a defect in the execution of the experiment within one square, that square can be omitted from the analysis.

4.1.2 Graeco-Latin squares

The Latin square design introduces a cross-classification of experimental units, represented by the rows and columns in the design arrangement, and a single set of treatments. From a combinatorial viewpoint there are three classifications on an equal footing, and we could, for example, write out the design labelling the new rows by the original treatments and inserting in the body of the square the original row numbering. In Chapter 6 we shall use the Latin square in this more symmetrical way.

We now discuss a development which is sometimes of direct use in applications but which is also important in connection with other designs. We introduce a further classification of the experimental units, also with v levels; it is convenient initially to denote the levels by letters of the Greek alphabet, using the Latin alphabet for the treatments. We require that the Greek letters also form a Latin square and further that each combination of a Latin and a Greek letter occurs together in the same cell just once. The resulting configuration is called a *Graeco-Latin square*. Combinatorially the four features, rows, columns and the two alphabets are on the same footing.

Illustration. In an industrial experiment comparing five treatments, suppose that the experiment is run for five days with five runs per day in sequence so that a 5 × 5 Latin square is a suitable design. Suppose now that five different but nominally similar machines are available for further processing of the material. It would then be reasonable to use the five further machines in a Latin square configuration and to require that each further machine is used once in combination with each treatment.

Table 4.1 shows an example of a Graeco-Latin square design before randomization, with treatments labelled $A, \ldots, E$ and machines $\alpha, \ldots, \epsilon$.

If the design is randomized as before, the analysis can be based on an assumed linear model or on randomization together with unit-treatment additivity. If $Y_{l_{st}g_{st}st}$ denotes the observation on row s, column t, Latin letter l_{st} and Greek letter g_{st} then the model that generates the results corresponding to randomization theory has

$$E(Y_{lgst}) = \mu + \tau_l + \nu_g + \beta_s + \gamma_t, \qquad (4.10)$$

where for simplicity we have abandoned the suffices on l and g.

Table 4.1 *A 5 × 5 Graeco-Latin square.*

$A\alpha$	$B\beta$	$C\gamma$	$D\delta$	$E\epsilon$
$B\gamma$	$C\delta$	$D\epsilon$	$E\alpha$	$A\beta$
$C\epsilon$	$D\alpha$	$E\beta$	$A\gamma$	$B\delta$
$D\beta$	$E\gamma$	$A\delta$	$B\epsilon$	$C\alpha$
$E\delta$	$A\epsilon$	$B\alpha$	$C\beta$	$D\gamma$

Again the model is formed in accordance with the principle that effects balanced out by design are represented by parameters in the model even though the effects may be of no intrinsic interest.

In the decomposition of the data vector the residual terms have the form
$$Y_{lgst} - \bar{Y}_{l...} - \bar{Y}_{.g..} - \bar{Y}_{..s.} - \bar{Y}_{...t} + 3\bar{Y}_{....}, \qquad (4.11)$$
and the sum of squares of these forms the residual sum of squares from which the error variance is estimated.

4.1.3 Orthogonal Latin squares

Occasionally it may be useful to add yet further alphabets to the above system. Moreover the system of arrangements that corresponds to such addition is of independent interest in connection with the generation of further designs.

It is not hard to show that for a $v \times v$ system there can be at most $(v-1)$ alphabets such that any pair of letters from two alphabets occur together just once. When v is a prime power, $v = p^m$, a complete set of $(v-1)$ orthogonal squares can be constructed; this result follows from some rather elegant Galois field theory briefly described in Appendix B. Table 4.2 shows as an example such a system for $v = 5$. It is convenient to abandon the use of letters and to label rows, columns and the letters of the various alphabets $0, \ldots, 4$. If such an arrangement, or parts of it, are used directly then rows, columns and the names of the letters of the alphabets would be randomized.

For the values of v likely to arise in applications, say $3 \leq v \leq 12$, a complete orthogonal set exists for the primes and prime powers, namely all numbers except 6, 10 and 12. Remarkably for $v = 6$ there does not exist even a Graeco-Latin square. For $v = 10$ and

Table 4.2 *Complete orthogonal set of* 5×5 *Latin squares.*

0000	1234	2413	3142	4321
1111	2340	3024	4203	0432
2222	3401	4130	0314	1043
3333	4012	0241	1420	2104
4444	0123	1302	2031	3210

12 such squares do exist: a pair of 10×10 orthogonal Latin squares was first constructed by Bose, Shrikhande and Parker (1960). At the time of writing it is not known whether a 10×10 square exists with more than two alphabets.

4.2 Incomplete block designs

4.2.1 General remarks

In the discussion so far the number of units per block has been assumed equal to the number of treatments, with an obvious extension if some treatments, for example a control, are to receive additional replication. Occasionally there may be some advantage in having the number of units per block equal to, say, twice the number of treatments, so that each treatment occurs twice in each block. A more common possibility, however, is that the number v of distinct treatments exceeds the most suitable choice for k, the number of units per block. This may happen because there is a firm constraint on the number of units per block, for example to $k = 2$ in a study involving twins. If blocks are formed on the basis of one day's work there will be some flexibility over the value of k, although ultimately an upper bound. In other cases there may be no firm restriction on k but the larger k the more heterogeneous the blocks and hence the greater the effective value of the error variance σ^2.

We therefore consider how a blocking system can be implemented when the number of units per block is less than v, the number of treatments. For simplicity we suppose that all treatments are replicated the same number r of times. The total number of experimental units is thus $n = rv$ and because this is also

INCOMPLETE BLOCK DESIGNS

bk, where b is the number of blocks, we have

$$rv = bk. \tag{4.12}$$

For given r, v, k it is necessary that b defined by this equation is an integer in order for a design to exist.

In this discussion we ignore any structure in the treatments, considering in particular all pairwise contrasts between $T_1, \ldots, T_v$ to be of equal interest.

One possible design would be to randomize the allocation subject only to equal replication, and to adjust for the resulting imbalance between blocks by fitting a linear model including block effects. This has, however, the danger of being quite inefficient, especially if subsets of treatments are particularly clustered together in blocks.

A better procedure is to arrange the treatments in as close to a balanced configuration as is achievable and it turns out that in a reasonable sense highest precision is achieved by arranging that each pair of treatments occurs together in the same block the same number, λ, say, of times. Since the number of units appearing in the same block as a given treatment, say T_1, can be calculated in two ways we have the identity

$$\lambda(v-1) = r(k-1). \tag{4.13}$$

Another necessary condition for the existence of a design is thus that this equation have an integer solution for λ. A design satisfying these conditions with $k < v$ is called a *balanced incomplete block design*.

A further general relation between the defining features is the inequality

$$b \geq v. \tag{4.14}$$

To see this, let N denote the $v \times b$ incidence matrix, which has entries n_{js} equal to 1 if the jth treatment appears in the sth block, and zero otherwise. Then

$$NN^T = (r - \lambda)I + \lambda 11^T, \tag{4.15}$$

where I is the identity matrix and 1 is a vector of unit elements, and it follows that NN^T and hence also N have rank v, but rank $N \leq \min(b, v)$, thus establishing (4.14).

Given values of r, v, b, k and λ satisfying the above conditions there is no general theorem to determine whether a corresponding balanced incomplete block design exists. The cases of practical in-

Table 4.3 *Existence of some balanced incomplete block designs.*

No. of Units per Block, k	No. of Treatments, v	No. of Blocks, b	No. of Replicates, r	Total no. of Units, $bk = rv$
2	3	3	2	6
2	4	6	3	12
2	5	10	4	20
2	any v	$\frac{1}{2}v(v-1)$	$(v-1)$	$v(v-1)$
3	4	4	3	12
3	5	10	6	30
3	6	10	5	30
3	6	20	10	60
3	7	7	3	21
3	9	12	4	36
3	10	30	9	90
3	13	26	6	78
3	15	35	7	105
4	5	5	5	20
4	6	15	10	60
4	7	7	4	28
4	8	14	7	56
4	9	18	8	72
4	10	15	6	60
4	13	13	4	52
4	16	20	5	80

terest have, however, been enumerated; see Table 4.3. Designs for two special cases are shown in Table 4.4, before randomization.

4.2.2 Construction

There is no general method of construction for balanced incomplete block designs even when they do exist. There are, however, some important classes of such design and we now describe just three.

INCOMPLETE BLOCK DESIGNS

Table 4.4 *Two special incomplete block designs. The first is resolvable into replicates I through VII.*

$k=3$	I	1 2 3	4 8 12	5 10 15	6 11 13	7 9 14
$v=15$	II	1 4 5	2 8 10	3 13 14	6 9 15	7 11 12
$b=35$	III	1 6 7	2 9 11	3 12 15	4 10 14	5 8 13
$r=7$	IV	1 8 9	2 13 15	3 4 7	5 11 4	6 10 12
	V	1 10 11	2 12 14	3 5 6	4 9 13	7 8 15
	VI	1 12 13	2 5 7	3 9 10	4 11 15	6 8 14
	VII	1 14 15	2 4 6	3 8 11	5 9 12	7 10 13

$k=4, v=9,$	1 2 3 4	1 2 5 6	1 2 7 8	1 3 5 7
$b=18, r=8$	1 4 6 8	1 3 6 9	1 4 8 9	1 5 7 9
	2 3 8 9	2 4 5 9	2 6 7 9	2 3 4 7
	2 5 6 8	3 5 8 9	4 6 7 9	3 4 5 6
	3 6 7 8	4 5 7 8		

The first are the so-called *unreduced designs* consisting of all combinations of the v treatments taken k at a time. The whole design can be replicated if necessary. The design has

$$b = \binom{v}{k}, \quad r = \binom{v-1}{k-1}. \tag{4.16}$$

Its usefulness is restricted to fairly small values of k, v such as in the paired design $k = 2$, $v = 5$ in Table 4.3.

A second family of designs is formed when the number of treatments is a perfect square, $v = k^2$, where k is the number of units per block, and a complete set of orthogonal $k \times k$ Latin squares is available, i.e. k is a prime power. We use the 5×5 squares set out in Table 4.2 as an illustration. We suppose that the treatments are set out in a key pattern in the form of a 5×5 square, namely

$$\begin{array}{ccccc} 1 & 2 & 3 & 4 & 5 \\ 6 & 7 & 8 & 9 & 10 \\ & & \vdots & & \end{array}$$

We now form blocks of size 5 by the following rules

1. produce 5 blocks each of size 5 via the rows of the key design
2. produce 5 more blocks each of size 5 via the columns of the key design
3. produce four more sets each of 5 blocks of size 5 via the four alphabets of the associated complete set of orthogonal Latin squares.

In general this construction produces the design with

$$v = k^2, \ r = k+1, \ b = k(k+1), \ \lambda = 1, \ n = k^2(k+1). \quad (4.17)$$

It has the special feature of resolvability, not possessed in general by balanced incomplete block designs: the blocks fall naturally into sets, each set containing each treatment just once. This feature is helpful if it is convenient to run each replicate separately, possibly even in different centres. Further it is possible, with minor extra complications, to analyze the design replicate by replicate, or omitting certain replicates. This can be a useful feature in protecting against mishaps occurring in some portions of the experiment, for example.

The third special class of balanced incomplete block designs are the symmetric designs in which

$$b = v, \ r = k. \quad (4.18)$$

Many of these can be constructed by numbering the treatments $0, \ldots, v-1$, finding a suitable initial block and generating the subsequent blocks by addition of 1 mod v. Thus with $v = 7$, $k = 3$ we may start with (0, 1, 3) and generate subsequent blocks as (1, 2, 4), (2, 3, 5), (3, 4, 6), (4, 5, 0), (5, 6, 1), (6, 0, 2), a design with $\lambda = 1$; see also Appendix B. Before use the design is to be randomized.

4.2.3 Youden squares

We introduced the $v \times v$ Latin square as a design for v treatments accommodating a cross-classification of experimental units so that variation is eliminated from error in two directions simultaneously. If now the number v of treatments exceeds the natural block sizes there are two types of incomplete analogue of the Latin square. In one it is possible that the full number v of rows is available, but there are only $k < v$ columns. It is then sensible to look for a design in which each treatment occurs just once in each column but the rows form a balanced incomplete block design. Such designs are

INCOMPLETE BLOCK DESIGNS

Table 4.5 *Youden square before randomization.*

0	1	2	3	4	5	6
1	2	3	4	5	6	0
3	4	5	6	0	1	2

called *Youden squares*; note though that as laid out the design is essentially rectangular not square.

Illustrations. Youden squares were introduced originally in connection with experiments in which an experimental unit is a leaf of a plant and there is systematic variation between different plants and between the position of the leaf on the plant. Thus with 7 treatments and 3 leaves per plant, taken at different positions down the plant, a design with each treatment occurring once in each position and every pair of treatments occurring together on the same plant the same number of times would be sensible. Another application is to experiments in which 7 objects are presented for scoring by ranking, it being practicable to look at 3 objects at each session, order of presentation within the session being relevant.

The incomplete block design formed by the rows has $b = v$, i.e. is symmetric and the construction at the end of the last subsection in fact in general yields a Youden square; see Table 4.5. In randomizing Table 4.5 as a Youden square the rows are permuted at random, then the columns are permuted at random holding the columns intact and finally the treatments are assigned at random to $(0, \ldots, v-1)$. If the design is randomized as a balanced incomplete block design the blocks are independently randomized, i.e. the column structure of the original construction is destroyed and the enforced balance which is the objective of the design disappears.

The second type of two-way incomplete structure is of less common practical interest and has fewer than v rows and columns. The most important of these arrangements are the *lattice squares* which have $v = k^2$ treatments laid out in $k \times k$ squares. For some details see Section 8.5.

4.2.4 Analysis of balanced incomplete block designs

We now consider the analysis of responses from a balanced incomplete block design; the extension to a Youden square is quite direct.

First it would be possible, and in a certain narrow sense valid, to ignore the balanced incomplete structure and to analyse the experiment as a completely randomized design or, in the case of a resolvable design, as a complete randomized block design regarding replicates as blocks. This follows from the assumption of unit-treatment additivity.

Nevertheless in nearly all contexts this would be a poor analysis, all the advantages of the blocking having been sacrificed and the effective error now including a component from variation between blocks. The exception might be if there were reasons after the experiment had been completed for thinking that the grouping into blocks had in fact been quite ineffective.

To exploit the special structure of the design which is intended to allow elimination from the treatment contrasts of systematic variation between blocks, we follow the general principle that effects eliminated by design are to be represented by parameters in the associated linear model. We suppose therefore that if treatment T_j occurs in block s the response is Y_{js}, where

$$E(Y_{js}) = \mu + \tau_j + \beta_s, \qquad (4.19)$$

and if it is necessary to resolve nonuniqueness of the parameterization we require by convention that

$$\Sigma \tau_j = \Sigma \beta_s = 0. \qquad (4.20)$$

Further we suppose that the Y_{js} are uncorrelated random variables of variance σ_k^2, using a notation to emphasize that the variance depends on the number k of units per block.

Because of the incomplete character of the design only those combinations of (j, s) specified by the $v \times b$ incidence matrix N of the design are observed. A discussion of the least squares estimation of the τ_j for general incomplete block designs is given in the next section; we outline here an argument from first principles. An alternative approach uses the method of fitting parameters in stages; see Appendix A2.6.

The right-hand side of the least squares equations consists of the

INCOMPLETE BLOCK DESIGNS

total of all observations, $Y_{..}$, and the treatment and block totals

$$S_j = \Sigma_{s=1}^{b} Y_{js} n_{js}, \quad B_s = \Sigma_{j=1}^{v} Y_{js} n_{js}. \quad (4.21)$$

In a randomized block design the τ_j can be estimated directly from the S_j but this is not so for an incomplete block design. We look for a linear combination of the (S_j, B_s) that is an unbiased estimate of, say, τ_j; this must be the least squares estimate required. The qualitative idea is that we have to "correct" S_j for the special features of the blocks in which T_j happens to occur.

Consider therefore the adjusted treatment total

$$Q_j = S_j - k^{-1} \Sigma_s n_{js} B_s; \quad (4.22)$$

note that the use of the incidence matrix ensures that the sum is over all blocks containing treatment T_j. A direct calculation shows that

$$E(Q_j) = r\tau_j + (r - \lambda)k^{-1} \Sigma_{l \neq j} \tau_l \quad (4.23)$$
$$= \frac{rv(k-1)}{k(v-1)} \tau_j, \quad (4.24)$$

where we have used the constraint $\Sigma \tau_j = 0$ and the identity defining λ. The least squares estimate of τ_j is therefore

$$\hat{\tau}_j = \frac{k(v-1)}{rv(k-1)} Q_j = \frac{k}{\lambda v} Q_j. \quad (4.25)$$

As usual we obtain an unbiased estimate of σ_k^2 from the analysis of variance table. Some care is needed in calculating this, because the treatment and block effects are not orthogonal. In the so-called *intrablock* or within block analysis, the sum of squares for treatments is adjusted for blocks, as in the computation of the least squares estimates above.

The sum of squares for treatments adjusted for blocks can most easily be computed by comparing the residual sum of squares after fitting the full model with that for the restricted model fitting only blocks. We can verify that the sum of squares due to treatment is $\Sigma_j \hat{\tau}_j^2 (\lambda v)/k = \Sigma_j Q_j^2 k/(\lambda v)$, giving the analysis of variance of Table 4.6.

For inference on a treatment contrast $\Sigma l_j \tau_j$, we use the estimate $\Sigma l_j \hat{\tau}_j$, which has variance

$$\operatorname{var}(\Sigma l_j \hat{\tau}_j) = \frac{k \sigma_k^2}{\lambda v} \Sigma l_j^2, \quad (4.26)$$

Table 4.6 *Intrablock analysis of variance for balanced incomplete block design.*

Source	Sum of squares	Degrees of freedom
Blocks (ignoring treatments)	$\Sigma_{j,s}(\bar{Y}_{.s} - \bar{Y}_{..})^2$	$b-1$
Treatments (adj. for blocks)	$k\Sigma_j Q_j^2/(\lambda v)$	$v-1$
Residual		$bk-v-b+1$
Total		$bk-1$

which is obtained by noting that $\mathrm{var}(Q_j) = r(k-1)\sigma_k^2/k$ and that, for $j \neq j'$, $\mathrm{cov}(Q_j, Q_{j'}) = -\lambda \sigma_k^2/k$. For example, if $l_1 = 1$, $l_2 = -1$, and $l_j = 0$, so that we are comparing treatments 1 and 2, we have $\mathrm{var}(\hat{\tau}_1 - \hat{\tau}_2) = 2k\sigma_k^2/(\lambda v)$. In the randomized block design with the same value of error variance, σ_k^2, and with r observations per treatment, we have $\mathrm{var}(\hat{\tau}_1 - \hat{\tau}_2) = 2\sigma_k^2/r$. The ratio of these

$$\mathcal{E} = \frac{v(k-1)}{(v-1)k}, \quad (4.27)$$

is called the *efficiency factor* of the design. Note that $\mathcal{E} < 1$. It essentially represents the loss of information incurred by having to unscramble the nonorthogonality of treatments and blocks.

To assess properly the efficiency of the design we must take account of the fact that the error variance in an ordinary randomized block design with blocks of v units is likely to be unequal to σ_k^2; instead the variance of the contrast comparing two treatments is

$$2\sigma_v^2/r \quad (4.28)$$

where σ_v^2 is the error variance when there are v units per block. The efficiency of the balanced incomplete block design is thus

$$\mathcal{E}\sigma_v^2/\sigma_k^2. \quad (4.29)$$

The smallest efficiency factor is obtained when $k = 2$ and v is large, in which case $\mathcal{E} = 1/2$. To justify the use of an incomplete block design we would therefore need $\sigma_k^2 \leq \sigma_v^2/2$.

The above analysis is based on regarding the block effects as arbitrary unknown parameters. The resulting intrablock least squares

INCOMPLETE BLOCK DESIGNS

analysis eliminates any effect of such block differences. The most unfavourable situation likely to arise in such an analysis is that in fact the blocking is ineffective, $\sigma_k^2 = \sigma_v^2$, and there is a loss of information represented by an inflation of variance by a factor $1/\mathcal{E}$.

Now while it would be unwise to use a balanced incomplete block design of low efficiency factor unless a substantial reduction in error variance is expected, the question arises as to whether one can recover the loss of information that would result in cases where the hoped-for reduction in error variance is in fact not achieved. Note in particular that if it could be recognized with confidence that the blocking was ineffective, the direct analysis ignoring the incomplete block structure would restore the variance of an estimated simple contrast to $2\sigma_v^2/r$.

A key to the recovery of information in general lies in the randomization applied to the ordering in blocks. This means that it is reasonable, in the absence of further information or structure in the blocks, to treat the block parameters β_s in the linear model as uncorrelated random variables of mean zero and variance, say σ_B^2. The resulting formulae can be justified by further appeal to randomization theory and unit-treatment additivity.

For a model-based approach we suppose that

$$Y_{js} = \mu + \tau_j + \beta_s + \epsilon_{js} \tag{4.30}$$

as before, but in addition to assuming ϵ_{js} to be independently normally distributed with zero mean and variance σ_k^2, we add the assumption that β_s are likewise independently normal with zero mean and variance σ_B^2. This can be justified by the randomization of the blocks within the whole experiment, or within replicates in the case of a resolvable design.

This model implies

$$B_s = Y_{.s} = k\mu + \Sigma_j n_{js}\tau_j + (k\beta_s + \Sigma_j n_{js}\epsilon_{js}), \tag{4.31}$$

where we might represent the error term in parentheses as η_s which has mean zero and variance $k(k\sigma_B^2 + \sigma_k^2)$. The least squares equations based on this model for the totals B_s give

$$\tilde{\mu} = \bar{B}_. = \bar{Y}_{..}, \tag{4.32}$$

$$\tilde{\tau}_j = \frac{\Sigma_s n_{js} Y_{.s} - rk\tilde{\mu}}{r - \lambda} \tag{4.33}$$

Table 4.7 *Interblock analysis of variance corresponding to Table 4.6.*

Source	Sum of squares	Degrees of freedom
Treatment (ignoring blocks)	$\Sigma_j(\bar{Y}_{j.} - \bar{Y}_{..})^2$	$v - 1$
Blocks (adj. for treatment)	by subtraction	$b - 1$
Residual	as in intrablock analysis	$bk - b - v + 1$

and it is easily verified that

$$\mathrm{var}(\tilde{\tau}_j) = \frac{k(v-1)}{v(r-\lambda)}(k\sigma_B^2 + \sigma_k^2). \qquad (4.34)$$

The expected value of the mean square due to blocks (adjusted for treatments) is $\sigma_k^2 + v(r-1)\sigma_B^2/(b-1)$ and unbiased estimates of the components of variance are available from the analysis of variance given in Table 4.7.

We have two sets of estimates of the treatment effects, $\hat{\tau}_j$ from the within block analysis and $\tilde{\tau}_j$ from the between block analysis. These two sets of estimates are by construction uncorrelated, so an estimate with smaller variance can be constructed as a weighted average of the two estimates, $w\hat{\tau}_j + (1-w)\tilde{\tau}_j$, where

$$w = \{\mathrm{var}(\hat{\tau}_j)\}^{-1}/[\{\mathrm{var}(\hat{\tau}_j)\}^{-1} + \{\mathrm{var}(\tilde{\tau}_j)\}^{-1}] \qquad (4.35)$$

will be estimated using the estimates of σ_k^2 and σ_B^2. This approach is almost equivalent to the use of a modified profile likelihood function for the contrasts under a normal theory formulation.

If σ_B^2 is large relative to σ_k^2, then the weight on $\hat{\tau}_j$ will be close to one, and there will be little gain in information over that from the intrablock analysis.

If the incomplete block design is resolvable, then we can remove a sum of squares due to replicates from the sum of squares due to blocks, so that σ_B^2 is now a component of variance between blocks within replicates.

4.2.5 More general incomplete block designs

There is a very extensive literature on incomplete block arrangements more general than balanced incomplete block designs. Some are simple modifications of the designs studied so far. We might, for example, wish to replicate some treatments more heavily than others, or, in a more extreme case, have one or more units in every block devoted to a control treatment. These and similar adaptations of the balanced incomplete block form are easily set up and analysed. In this sense balanced incomplete block designs are as or more important for constructing designs adapted to the special needs of particular situations as they are as designs in their own right.

Another type of application arises when no balanced incomplete block design or *ad hoc* modification is available and minor changes in the defining constants, v, r, b, k, are unsatisfactory. Then various types of partially balanced designs are possible. Next in importance to the balanced incomplete block designs are the *group divisible designs*. In these the v treatments are divided into a number of equal-sized disjoint sets. Each treatment is replicated the same number of times and appears usually either once or not at all in each block. Slightly more generally there is the possibility that each treatment occurs either $[v/k]$ or $[v/k+1]$ times in each block, where there are k units per block. The association between treatments is determined by two numbers λ_1 and λ_2. Any two treatments in the same set occur together in the same block λ_1 times whereas any two treatments in different sets occur together λ_2 times. We discuss the role of balance a little more in Chapter 7 but in essence balance forces good properties on the eigenvalues of the matrix C, defined at (4.39) below, and hence on the covariance matrix of the estimated contrasts.

Sometimes, for example in plant breeding trials, there is a strong practical argument for considering only resolvable designs. It is therefore useful to have a general flexible family of such designs capable of adjusting to a range of requirements. Such a family is defined by the so-called α *designs*; the family is sufficiently rich that computer search within it is often needed to determine an optimum or close-to-optimum design. See Exercise 4.6.

While quite general incomplete block designs can be readily analysed as a linear regression model of the type discussed in Appendix A, the form of the intrablock estimates and analysis of variance is

Table 4.8 *Analysis of variance for a general incomplete block design.*

Source	Sum of squares	Degrees of freedom
Blocks (ignoring treatments)	$B^T K^{-1} B - Y_{...}^2/n$	$b - 1$
Trt (adj. for blocks)	$Q^T \hat{\tau}$	$v - 1$
Residual	by subtraction	$n - b - v + 1$
Total	$\Sigma Y_{jsm}^2 - Y_{...}^2/n$	$n - 1$

still relatively simple and we briefly outline it now. We suppose that treatment j is replicated r_j times, that block s has k_s experimental units, and that the jth treatment appears in the sth block n_{js} times. The model for the mth observation of treatment j in block s is

$$Y_{jsm} = \mu + \tau_j + \beta_s + \epsilon_{jsm}. \qquad (4.36)$$

The adjusted treatment totals are

$$Q_j = S_j - \Sigma_s n_{js} B_s / k_s, \qquad (4.37)$$

where, as before, S_j is the total response on the jth treatment and B_s is the total for the sth block. We have again

$$E(Q_j) = r_j \tau_j - \Sigma_l (\Sigma_s n_{js} n_{ls}/k_s) \tau_l \qquad (4.38)$$

or, defining $Q = \text{diag}(Q_1, \ldots, Q_v)$, $R = \text{diag}(r_1, \ldots, r_v)$, $K = \text{diag}(k_1, \ldots, k_b)$ and $N = ((n_{js}))$,

$$E(Q) = (R - NK^{-1}N^T)\tau = C\tau, \qquad (4.39)$$

say, and hence

$$Q = C\hat{\tau}. \qquad (4.40)$$

The $v \times v$ matrix C is not of full rank, but if every pair of contrasts $\tau_j - \tau_l$ is to be estimable, the matrix must have rank $v - 1$. Such a design is called *connected*. We may impose the constraint $\Sigma \tau_j = 0$ or $\Sigma r_j \tau_j = 0$, leading to different least squares estimates of τ but the same estimates of contrasts and the same analysis of variance table. The analysis of variance is outlined in Table 4.8, in which we write B for the vector of block totals.

The general results for the linear model outlined in Appendix A

INCOMPLETE BLOCK DESIGNS

can be used to show that the covariance matrix for the adjusted treatment totals is given by

$$\text{cov}(Q) = (R - NK^{-1}N^T)\sigma^2, \tag{4.41}$$

where σ^2 is the variance of a single response. This leads directly to an estimate of the variance of any linear contrast $\Sigma l_i \hat{\tau}_i$ as $l^T C^- l$, where a specific form for the generalized inverse, C^-, can be obtained by invoking either of the constraints $1^T \tau = \Sigma \tau_j = 0$ or $1^T R \tau = \Sigma r_j \tau_j = 0$.

These formulae can be used not only in the direct analysis of data but also to assess the properties of nonstandard designs constructed from *ad hoc* considerations. For example if no balanced incomplete block design exists but another design can be shown to have variance properties close to those that a balanced incomplete block design would have had then the new design is likely to be close to optimal. Further the relative efficiencies can be compared using the appropriate C^- for each design.

4.2.6 Examples

We first discuss a simple balanced incomplete block design.

Example I of Cox and Snell (1981) is based on an experiment reported by Biggers and Heyner (1961) on the growth of bones from chick embryos after cultivation over a nutrient medium. There were two bones available from each embryo, and each embryo formed a block. There were six treatments, representing a complete medium and five other media obtained by omitting a single amino acid. The design and data are given in Table 4.9. The treatment assignment was randomized, but the data are reported in systematic order. In the notation of the previous subsection, $k = 2$, $\lambda = 1$, $v = 6$ and $r = 5$.

The raw treatment means, and the means adjusted for blocks, are given in Table 4.10, and these form the basis for the intrablock analysis. The intrablock analysis of variance table is given in Table 4.11, from which an estimate of σ is 0.0811.

The treatment effect estimates $\hat{\tau}_j$, under the constraint $\Sigma \tau_j = 0$, are given by $2Q_j/6$, and the standard error for the difference between any two $\hat{\tau}_j$ is $\sqrt{(4/6)} \times 0.0811 = 0.066$.

These estimates can be combined with the interblock estimates, which are obtained by fitting a regression model to the block totals. The intrablock and interblock effect estimates are shown in

Table 4.9 *Log dry weight (µg) of chick bones for 15 embryos.*

1	C	2.51	His-	2.15	9	His-	2.32	Lys-	2.53
2	C	2.49	Arg-	2.23	10	Arg-	2.15	Thr-	2.23
3	C	2.54	Thr-	2.26	11	Arg-	2.34	Val-	2.15
4	C	2.58	Val-	2.15	12	Arg-	2.30	Lys-	2.49
5	C	2.65	Lys-	2.41	13	Thr-	2.20	Val-	2.18
6	His-	2.11	Arg-	1.90	14	Thr-	2.26	Lys-	2.43
7	His-	2.28	Thr-	2.11	15	Val-	2.28	Lys-	2.56
8	His-	2.15	Val-	1.70					

Table 4.10 *Adjusted and unadjusted treatment means of log dry weight.*

	C	His-	Arg-	Thr-	Val-	Lys-
$\bar{Y}_{j.}$ (Unadj. mean)	2.554	2.202	2.184	2.212	2.092	2.484
$\hat{\tau}_j + \bar{Y}_{..}$ (Adj. mean)	2.550	2.331	2.196	2.201	2.060	2.390

Table 4.11 *Analysis of variance of log dry weight.*

Source	Sum of sq.	D.f.	Mean sq.
Days	0.7529	14	0.0892
Treatments (adj.)	0.4462	5	0.0538
Residual	0.0658	10	0.0066

Table 4.12 *Within and between block estimates of treatment effects.*

	C	His-	Arg-	Thr-	Val-	Lys-
$\hat{\tau}_j$	0.262	0.043	−0.092	−0.087	−0.228	0.102
$\tilde{\tau}_j$	0.272	−0.280	−0.122	−0.060	−0.148	0.338
τ_j^*	0.264	−0.017	−0.097	−0.082	−0.213	0.146

Table 4.12, along with the pooled estimate τ_j^*, which is a linear combination weighted by the inverse variances. Because there is relatively large variation between blocks, the pooled estimates are not very different from the within block estimates. The standard error for the difference between two τ_j^*'s is 0.059.

As a second example we take an incomplete block design that was used in a study of the effects of process variables on the properties of pastry dough. There were 15 different treatments, and the experiment took seven days. It was possible to do just four runs on each day. The data are given in Table 4.13. The response given here is the cross-sectional expansion index for the pastry dough, in cm per g. The treatments have a factorial structure, which is ignored in the present analysis. The treatment means and adjusted treatment means are given in Table 4.14.

Note that all the treatments used on Day 6, the day with the highest block mean, were also used on other days, whereas three of the treatments used on Day 5 were not replicated. Thus there is considerable intermixing of block effects with treatment effects.

The analysis of variance, with treatments adjusted for blocks, is given in Table 4.15. It is mainly useful for providing an estimate of variance for comparing adjusted treatment means. As the detailed treatment structure has been ignored in this analysis, we defer the comparison of treatment means to Exercise 6.9.

4.3 Cross-over designs

4.3.1 General remarks

One of the key assumptions underlying all the previous discussion is that the response on any unit depends on the treatment applied to that unit independently of the allocation of treatments to the

Table 4.13 *An unbalanced incomplete block design. From Gilmour and Ringrose (1999). 15 treatments; response is expansion index of pastry dough (cm per g).*

					Mean
	1	8	9	9	
Day 1	15.0	14.8	13.0	11.7	13.625
	9	5	4	9	
Day 2	12.2	14.1	11.2	11.6	12.275
	2	3	8	5	
Day 3	15.9	10.8	15.8	15.6	14.525
	12	6	14	10	
Day 4	12.7	18.6	11.4	11.2	13.475
	11	15	3	13	
Day 5	13.0	11.1	10.1	11.7	11.475
	1	6	4	7	
Day 6	14.6	17.8	12.8	15.4	15.1
	2	9	7	9	
Day 7	15.0	10.7	10.9	9.6	11.55

Table 4.14 *Treatment means of expansion index: unadjusted and adjusted for days.*

Trt	1	2	3	4	5	6	7	8
$\bar{Y}_{j.}$	14.8	15.4	10.4	12.0	14.9	18.2	13.1	15.3
$\hat{\tau}_j + \bar{Y}_{..}$	14.5	16.7	10.8	12.1	15.3	17.1	14.0	15.4

Trt	9	10	11	12	13	14	15
$\bar{Y}_{j.}$	11.5	11.2	13.0	12.7	11.7	11.4	11.1
$\hat{\tau}_j + \bar{Y}_{..}$	12.6	9.7	13.7	11.2	12.4	9.9	11.8

Table 4.15 *Within-block analysis of variance for pastry example.*

Source	Sum of sq.	D. f.	Mean sq.
Days (ign. trt)	49.41	6	8.235
Treatment (adj. for days)	96.22	14	6.873
Residuals	5.15	7	0.736

other units. When the experimental units are physically different entities this assumption will, with suitable precautions, be entirely reasonable.

Illustration. In an agricultural fertiliser trial in which the experimental units are distinct plots of ground the provision of suitable guard rows between the plots will be adequate security against diffusion of fertiliser from one plot to another affecting the results.

When, however, the same physical object or individual is used as an experimental unit several times the assumption needs more critical attention and may indeed be violated.

Illustrations. In a typical investigation in experimental psychology, subjects are exposed to a series of conditions (treatments) and appropriate responses to each observed; a Latin square design may be very suitable. It is possible, however, in some contexts that the response in one period is influenced not only by the condition in that period but by that in the previous period and perhaps even on the whole sequence of conditions encountered up to that point. This possibility would distort and possibly totally vitiate an interpretation based on the standard Latin square analysis. We shall see that if the effect of previous conditions is confined to a simple effect one period back, then suitable designs are available. If, however, there is the possibility of more complex forms of dependence on previous conditions, it will be preferable to reconsider the whole basis of the design, perhaps exposing each subject to only one experimental condition or perhaps regarding a whole sequence of conditions as defining a treatment.

An illustration of a rather different kind is to so-called community intervention trials. Here a whole village, community or school is an experimental unit, for example to compare the effects of health education campaigns. Each school, say, is randomized to one of a number of campaigns. Here there can be interference between units in two different senses. First, unless the schools are far apart, it may be difficult to prevent children in one school learning what is happening in other schools. Secondly there may be migration of children from one school to another in the middle of the programme under study. To which regime should such children be assigned? If the latter problem happens only on a small scale it can be dealt with in analysis, for example by omitting the children in question. Thirdly there is the delicate issue of what information can be learned from the variation between children within schools.

In experiments on the effect of diet on the milk yield of cows a commonly used design is again a Latin square in which over the period of lactation each cow receives each of a number of diets. Similar remarks about carry-over of treatment effects apply as in the psychological application.

In an experiment on processing in the wool textile industries one set of treatments may correspond to different amounts of oil applied to the raw material. It is possible that some of the oil is retained on the machinery and that a batch processed following a batch with a high oil allocation in effect receives an additional supplement of oil.

Finally a quite common application is in clinical trials of drugs. In its simplest form there are two treatments, say T and C; some patients receive the drug T in the first period and C in the second period, whereas other patients have the order C, T. This, the two-treatment, two-period design can be generalized in various obvious ways by extending the number of treatments or the number of periods or both; see Exercise 1.2.

We call any effect on one experimental unit arising from treatments applied to another unit a *carry-over* or *residual* effect. It is unlikely although not impossible that the carry-over of a treatment effect from one period to another is of intrinsic interest. For the remainder of the discussion we shall, however, take the more usual view that any such effects are an encumbrance.

Two important general points are that even in the absence of carry-over effects it is possible that treatment effects estimated in an environment of change are not the same as those in a more stable context. For example, cows subject to frequently changing diets might react differently to a specific diet from how they would

CROSS-OVER DESIGNS 89

under a stable environment. If that seems a serious possibility the use of physically the same material as a unit several times is suspect and alternative designs should be explored.

The second point is that it will often be possible to eliminate or at least substantially reduce carry-over effects by wash-out periods restoring the material so far as feasible to its initial state. For example, in the textile experiment mentioned above the processing of each experimental batch could be preceded by a standard batch with a fixed level of oil returning the machinery to a standard state. Similar possibilities are available for the dairy and pharmaceutical illustrations.

Such wash-out periods are important, their duration being determined from subject-matter knowledge, although unless they can be made to yield further useful information the effort attached to them may detract from the appeal of this type of design.

In some agricultural and ecological applications, interference originates from spatial rather than temporal contiguity. The design principles remain the same although the details are different. The simplest possibility is that the response on one unit depends not only on the treatment applied to that unit but also on the treatment applied to spatially adjacent units; sometimes the dependence is directly on the responses from the other units. These suggest the use of designs in which each treatment is a neighbour of each other treatment approximately the same number of times. A second possibility arises when, for example, tall growing crops shield nearby plots from sunlight; in such a case the treatments might be varieties growing to different heights. To some extent this can be dealt with by sorting the varieties into a small number of height groups on the basis of *a priori* information and aiming by design to keep varieties in the same group nearby as far as feasible.

In summary, repeated use of the same material as an experimental unit is probably wise only when there are solid subject-matter arguments for supposing that any carry-over effects are either absent or if present are small and of simple form. Then it is often a powerful technique.

4.3.2 Two-treatment two-period design

We begin with a fairly thorough discussion of the simplest design with two treatments and two periods, in effect an extension of the discussion of the matched pair design in Section 3.3. We regard the two periods asymmetrically, supposing that all individuals start the first period in the same state whereas that is manifestly not true for the second period. In the first period let the treatment parameter be $\pm\delta$ depending on whether T or C is used and in the second period let the corresponding parameter be $\pm(\delta+\gamma)$, where γ measures a treatment by period interaction. Next suppose that in the second period there is added to the treatment effect a carry-over or residual effect of $\pm\rho$ following T or C, respectively, in the first period. Finally let π denote a systematic difference between observations in the second period as compared with the first.

If an individual additive parameter is inserted for each pair of units, i.e. for each patient in the clinical trial context, an analysis free of these parameters must be based on differences of the observation in the second period minus that in the first. If, however, as would typically be the case, individuals have been randomly allocated to a treatment order a second analysis is possible based on the pair totals. The error for this will include a contribution from the variation between individuals and so the analysis of the totals will have a precision lower than and possibly very much lower than that of the analysis of differences. This is analogous to the situation arising in interblock and intrablock analyses of the balanced incomplete block design.

For individuals receiving treatments in the order C, T the expected value of the first and second observations are, omitting individual unit effects,

$$\mu - \delta, \quad \mu + \delta + \gamma - \rho + \pi, \tag{4.42}$$

whereas for the complementary sequence the values are

$$\mu + \delta, \quad \mu - \delta - \gamma + \rho + \pi. \tag{4.43}$$

Thus the mean of the differences within individuals estimates respectively

$$2\delta + \pi + \gamma - \rho, \quad -2\delta + \pi - \gamma + \rho. \tag{4.44}$$

and the difference of these estimates leads to the estimation of

$$2\Delta + 2(\gamma - \rho), \tag{4.45}$$

where $\Delta = 2\delta$. On the other hand a similar comparison based on the sums of pairs of observations leads to the estimation, typically with low precision, of $2(\gamma - \rho)$.

From this we can see that treatment by period interaction and

CROSS-OVER DESIGNS 91

residual effect are indistinguishable. In addition, assuming that the first period treatment effect Δ is a suitable target parameter, it can be estimated from the first period observations alone, but only with low precision because the error includes a between individual component. If this component of variance were small there would be little advantage to the cross-over design anyway.

This analysis confirms the general conclusion that the design is suitable only when there are strong subject-matter arguments for supposing that any carry-over effect or treatment by period interaction are small and that there are substantial systematic variations between subjects.

To a limited extent the situation can be clarified by including some individuals receiving the same treatment in both periods, i.e. T, T or C, C. Comparison of the differences then estimates $(\gamma + \rho)$, allowing the separation of treatment by period interaction from residual effect. Comparison of sums of pairs leads to the estimation of $2\Delta + (\gamma + \rho)$, typically with low precision. Note that this assumes the absence of a direct effect by carry-over effect interaction, i.e. that the carry-over effect when there is no treatment change is the same as when there is a change. The reasonableness of this assumption depends entirely on the context.

Another possibility even with just two treatments is to have a third period, randomizing individuals between the six sequences

$$T,T,C;\ T,C,T;\ T,C,C;\ C,C,T;\ C,T,C;\ C,T,T. \qquad (4.46)$$

It might often be reasonable to assume the absence of a treatment by period interaction and to assume that the carry-over parameter from period 1 to period 2 is the same as, say, between period 2 and period 3.

4.3.3 Special Latin squares

With a modest number of treatments the Latin square design with rows representing individuals and columns representing periods is a natural starting point and indeed many of the illustrations of the Latin square mentioned in Chapter 4 are of this form, ignoring, however, carry-over effects and treatment by period interactions.

The general comments made above about the importance of wash-out periods and the desirability of restricting use of the designs to situations where any carry-over effects are of simple form and small continue to hold. If, however, the possibility of treatment by period interaction is ignored and any carry-over effect is assumed to be an additive effect extending one period only, it is appealing to look for special Latin square designs in which each treatment follows each other treatment the same number of times, either over each square, or, if this is not possible over the set of Latin squares forming the whole design.

We first show how to construct such special squares. For a $v \times v$ square it is convenient to label the treatments $0, \ldots, v-1$ without any implication that the treatments correspond to levels of a quantitative variable.

If v is even the key to design construction is that the sequence

$$0, 1, v-1, 2, v-2 \ldots \quad (4.47)$$

has all possible differences mod v occurring just once. Thus if we generate a Latin square from the above first row by successive addition of 1 followed by reduction mod v it will have the required properties. For example, with $v = 6$ we get the following design in which each treatment follows each other treatment just once:

$$\begin{array}{cccccc} 0 & 1 & 5 & 2 & 4 & 3 \\ 1 & 2 & 0 & 3 & 5 & 4 \\ 2 & 3 & 1 & 4 & 0 & 5 \\ 3 & 4 & 2 & 5 & 1 & 0 \\ 4 & 5 & 3 & 0 & 2 & 1 \\ 5 & 0 & 4 & 1 & 3 & 2 \end{array}$$

The whole design might consist of several such squares independently randomized by rows and by naming the treatments but not, of course, randomized by columns.

If v is odd the corresponding construction requires pairs of squares generated by the first row as above and by that row reversed.

Note that in these squares no treatment follows itself so that in the presence of simple carry-over effects the standard Latin square analysis ignoring carry-over effects yields biased estimates of the direct treatment effects. This could be obviated if every row is preceded by a period with no observation but in which the same treatment is used as in the first period of the standard square. It is unlikely that this will often be a very practicable possibility.

For continuous observations the analysis of such a design would usually be via an assumed linear model in which if Y_{suij} is the observation in row s, column or period u receiving treatment i and

CROSS-OVER DESIGNS

previous treatment j then
$$E(Y_{suij}) = \mu + \alpha_s + \beta_u + \tau_i + \rho_j, \tag{4.48}$$
with the usual assumptions about error and conventions to define the parameters in the overparameterized model, such as
$$\Sigma\alpha_s = \Sigma\beta_u = \Sigma\tau_i = \Sigma\rho_j = 0. \tag{4.49}$$
Note also that any carry-over effect associated with the state of the individual before the first treatment period can be absorbed into β_0, the first period effect.

In practice data from such a design are usually most simply analysed via some general purpose procedure for linear models. Nevertheless it is helpful to sketch an approach from first principles, partly to show what is essentially involved, and partly to show how to tackle similar problems in order to assess the precision likely to be achieved.

To obtain the least squares estimates, say from a single square with an even value of v, we argue as follows. By convention number the rows so that row s ends with treatment s. Then the least squares equations associated with general mean, row, column, direct and carry-over treatment effects are respectively

$$v^2\hat{\mu} = Y_{....}, \tag{4.50}$$
$$v\hat{\mu} + v\hat{\alpha}_s - \hat{\rho}_s = Y_{s...}, \tag{4.51}$$
$$v\hat{\mu} + v\hat{\beta}_u = Y_{.u..}, \tag{4.52}$$
$$v\hat{\mu} + v\hat{\tau}_i - \hat{\rho}_i = Y_{..i.}, \tag{4.53}$$
$$(v-1)\hat{\mu} - \hat{\tau}_j + (v-1)\hat{\rho}_j - \hat{\beta}_1 = Y_{...j}. \tag{4.54}$$

It follows that
$$(v^2 - v - 1)\hat{\tau}_i = \{(v-1)Y_{..i.} + Y_{...i}\}, \tag{4.55}$$
although to satisfy the standard constraint $\Sigma\tau_i = 0$ a constant $-Y_{....} + Y_{.1..}/v$ has to be added to the right-hand side, not affecting contrasts.

A direct calculation now shows that the variance of a simple contrast is
$$\operatorname{var}(\hat{\tau}_1 - \hat{\tau}_2) = 2\sigma^2(v-1)/(v^2 - v - 1) \tag{4.56}$$
showing that as compared with the variance $2\sigma^2/v$ for a simple comparison of means the loss of efficiency from nonorthogonality is small.

This analysis assumes that the parameters of primary interest are contrasts of the direct treatment effects, i.e. contrasts of the τ_i. If, however, it is reasonable to assume that the carry-over effect of a treatment following itself is the same as following any other treatment then the total treatment effect $\tau_i + \rho_i$ would be the focus of attention.

In some contexts also a more sensitive analysis would be obtained if it were reasonable to make the working assumption that $\rho_i = \kappa \tau_i$ for some unknown constant of proportionality κ, leading to a nonlinear least squares analysis. There are many extensions of these ideas that are theoretically possible.

4.3.4 Single sequence

In the above discussion it has been assumed that there are a number of individuals on each of which observation continues for a relatively modest number of periods. Occasionally, however, there is a single individual under study and then the requirement is for a single sequence possibly quite long, in which each treatment occurs the same number of times and, preferably in which some balance of carry-over effects is achieved.

Illustration. Most clinical trials involve quite large and sometimes very large numbers of patients and are looking for relatively small effects in a context in which treatment by patient interaction is likely to be small, i.e. any treatment effect is provisionally assumed uniform across patients. By contrast it may happen that the search is for the treatment appropriate for a particular patient in a situation in which it is possible to try several treatments until, hopefully, a satisfactory solution is achieved. Such designs are called in the literature *n of one* designs. If the trial is over an extended period we may need special sequences of the type mentioned above.

The need for longish sequences of a small number of letters in which each letter occurs the same number of times and the same number of times either following all other letters or all letters, i.e. including itself, will be discussed in the context of serially correlated errors in Section 8.4. In the context of n of one trials it may be that the objective is best formulated not as estimating treatment contrasts but rather in decision-theoretic terms as that of maintaining the patient in a satisfactory condition for as high a proportion of time as possible. In that case quite different strategies may be appropriate, broadly of the play-the-winner kind, in which successful treatments are continued until failure when they are replaced either from a pool of as-yet-untried treatments or by the apparently best of the previously used and temporarily abandoned treatments.

4.4 Bibliographic notes

Latin squares, studied by Euler for their combinatorial interest, were occasionally used in experimentation in the 19th century. Their systematic study and the introduction of randomization is due to R. A. Fisher. For a very detailed study of Latin squares, including an account of the chequered history of the 6×6 Graeco-Latin square, see Denés and Keedwell (1974). For an account of special forms of Latin squares and of various extensions, such as to Latin cubes, see Preece (1983, 1988).

Balanced incomplete block designs were introduced by Yates (1936). The extremely extensive literature on their properties and extensions is best approached via P. W. M. John (1971); see also J. A. John and E. R. Williams (1995) and Raghavarao (1971). Hartley and Smith (1948) showed by direct arguments that every symmetric balanced incomplete block design, i.e. one having $b = v$, can be converted into a Youden square by suitable reordering. For a careful discussion of efficiency in incomplete block designs, see Pearce (1970) and related papers.

Cross-over designs are discussed in books by Jones and Kenward (1989) and Senn (1993), the latter largely in a clinical trial context. The special Latin squares balanced for residual effects were introduced by E. J. Williams (1949, 1950). For some generalizations of Williams's squares using just three squares in all, see Newcombe (1996). An analysis similar to that in Section 4.3.3 can be found in Cochran and Cox (1958, Section 4.6a). For some of the problems encountered with missing data in cross-over trials, see Chao and Shao (1997).

For n of one designs, see Guyatt et al. (1986).

4.5 Further results and exercises

1. Show that randomizing a Latin square by (a) permuting rows and columns at random, (b) permuting rows, columns and treat-

ment names at random, (c) choosing at random from the set of all possible Latin squares of the appropriate size all generate first and second moment randomization distributions with the usual properties. What considerations might be used to choose between (a), (b) and (c)? Explore a low order case numerically.

2. Set out an orthogonal partition of 6×6 Latin squares in which a second alphabet is superimposed on the first with two letters each occurring three times in each row and column and three times coincident with each letter of the Latin square. Find also an arrangement with three letters each occurring twice in each row, etc. Give the form of the analysis of variance in each case. These partitions in the 6×6 case were given by Finney (1945a) as in one sense the nearest one could get to a 6×6 Graeco-Latin square.

3. In 1850 the Reverend T.P. Kirkman posed the following problem. A schoolmistress has a class of 15 girls and wishes to take them on a walk each weekday for a week. The girls should walk in threes with no two girls together in a triplet more than once in the week. Show that this is equivalent to finding a balanced incomplete block design and give an explicit solution.

4. An alternative to a balanced incomplete block design is to use a control treatment within each block and to base the analysis on differences from the control. Let there be v treatments to be compared in b blocks each of k units, of which c units receive the control C. Suppose for simplicity that $v = b(k - c)$, and that the design consists of r replicates each of b blocks in each of which every treatment occurs once and the control bc times. Discuss the advantages of basing the analysis on (a) the difference between each treated observation and the mean of the corresponding controls and (b) on an analysis of covariance, treating the mean of the control observations as an explanatory variable. For (a) and for (b) calculate the efficiency of such a design relative to a balanced incomplete block design without recovery of interblock information and using no observations on control and with respect to a completely randomized design ignoring block structure. How would the conclusions be affected if C were of intrinsic interest and not merely a device for improving precision? The approach was discussed by Yates (1936) who considered it inefficient. His conclusions were confirmed by

FURTHER RESULTS AND EXERCISES

Atiqullah and Cox (1962) whose paper has detailed theoretical calculations.

5. Develop the intra-block analysis of an incomplete block design for v ordinary treatments, and one extra treatment, a control, in which the ordinary treatments are to be set out in a balanced incomplete block design and the control is to occur on k_c extra units in each block. Find the efficiency factors for the comparison of two ordinary treatments and for the comparison of an ordinary treatment with control. Show how to choose k_c to optimize the comparison of the v ordinary treatments individually with the control.

6. The family of resolvable designs called $\alpha(h_1, \ldots, h_q)$ designs for $v = wk$ treatments, r replicates of each treatment, k units per block and with b blocks in each replicate are generated as follows. There are r resolution classes of w blocks each. Any pair of treatments occurs together in h_1 or ... or in h_q blocks. In many ways the most important cases are the $\alpha(0, 1)$ designs followed by the $\alpha(0, 1, 2)$ designs; that is, in the former case each pair of treatments occurs together in a block at most once.

    The method of generation starts with an initial block within each resolution class, and generates the blocks in that resolution class by cyclic permutation.

    These designs were introduced by Patterson and Williams (1976) and have been extensively used in plant breeding trials in which large numbers of varieties may be involved with a fairly small number of replicates of each variety and in which resolvability is important on practical grounds. See John and Williams (1995) for a careful discussion within a much larger setting and Street and Street (1987) for an account with more emphasis on the purely combinatorial aspects. For resolvable designs with unequal block sizes, see John et al. (1999).

7. Raghavarao and Zhou (1997, 1998) have studied incomplete block designs in which each triple of treatments occur together in a block the same number of times. One application is to marketing studies in which v versions of a commodity are to be compared and each shop is to hold only $k < v$ versions. The design ensures that each version is seen in comparison with each pair of other possibilities the same number of times. Give such a design for $v = 6, k = 4$ and suggest how the responses might be analysed.

CHAPTER 5

Factorial designs: basic ideas

5.1 General remarks

In previous chapters we have emphasized experiments with a single unstructured set of treatments. It is very often required to investigate the effect of several different sets of treatments, or more generally several different explanatory factors, on a response of interest. Examples include studying the effect of temperature, concentration, and pressure on the hardness of a manufactured product, or the effects of three different types of fertiliser, say nitrogen, potassium and potash, on the yield of a crop. The different aspects defining treatments are conventionally called *factors*, and there are typically a specified, usually small, number of *levels* for each factor. An individual treatment is a particular combination of levels of the factors.

A *complete factorial experiment* consists of an equal number of replicates of all possible combinations of the levels of the factors. For example, if there are three levels of temperature, and two each of concentration and pressure, then there are $3 \times 2 \times 2 = 12$ treatments, so that we will need at least 12 experimental units in order to study each treatment only once, and at least 24 in order to get an independent estimate of error from a complete replicate of the experiment.

There are several reasons for designing complete factorial experiments, rather than, for example, using a series of experiments investigating one factor at a time. The first is that factorial experiments are much more efficient for estimating main effects, which are the averaged effects of a single factor over all units. The second, and very important, reason is that interaction among factors can be assessed in a factorial experiment but not from series of one-at-a-time experiments.

Interaction effects are important in determining how the conclusions of the experiment might apply more generally. For example, knowing that nitrogen only improves yield in the presence of potash would be crucial information for general recommendations on fertiliser usage. A main basis for empirical extrapolation of conclusions is demonstration of the absence of important interactions. In other contexts interaction may give insight into how the treatments "work". In many medical contexts, such as recently developed treatments for AIDS, combinations of drugs are effective when treatment with individual drugs is not.

Complete factorial systems are often large, especially if an appreciable number of factors is to be tested. Often an initial experiment will set each factor at just two levels, so that important main effects and interactions can be quickly identified and explored further. More generally a balanced portion or fraction of the complete factorial can often be used to get information on the main effects and interactions of most interest.

The choice of factors and the choice of levels for each factor are crucial aspects of the design of any factorial experiment, and will be dictated by subject matter knowledge and constraints of time or cost on the experiment.

The levels of factors can be qualitative or quantitative. Quantitative factors are usually constructed from underlying continuous variables, such as temperature, concentration, or dose, and there may well be interest in the shape of the response curve or response surface. Factorial experiments are an important ingredient in response surface methods discussed further in Section 6.5. Qualitative factors typically have no numerical ordering, although occasionally factors will have a notion of rank that is not strictly quantitative.

Factors are initially thought of as aspects of treatments: the assignment of a factor level to a particular experimental unit is under the investigator's control and in principle any unit might receive any of the various factor combinations under consideration.

For some purposes of design and analysis, although certainly not for final interpretation, it is helpful to extend the definition of a factor to include characteristics of the experimental units. These may be either important intrinsic features, such as sex or initial body mass, or nonspecific aspects, such as sets of apparatus, centres in a clinical trial, etc. stratifying the experimental units.

Illustrations. In a laboratory experiment using mice it might often be reasonable to treat sex as a formal factor and to ensure that each treatment factor occurs equally often with males and

EXAMPLE

females. In an agricultural field trial it will often be important to replicate the experiment, preferably in virtually identical form, in a number of farms. This gives sex in the first case and farms in the second some of the features of a factor. The objective is not to compare male and female mice or to compare farms but rather to see whether the conclusions about treatments differ for male and for female mice or whether the conclusions have a broad range of validity across different farms.

As noted in Section 3.2 for analysis and interpretation it is often desirable to distinguish between specific characteristics of the experimental units and nonspecific groupings of the units, for example defining blocks in a randomized block experiment.

For most of the subsequent discussion we take the factors as defining treatments. Regarding each factor combination as a treatment, the discussion of Chapters 3 and 4 on control of haphazard error applies, and we may, for example, choose a completely randomized experiment, a randomized block design, a Latin square, and so on. Sometimes replication of the experiment will be associated with a blocking factor such as days, laboratories, etc.

5.2 Example

This example is adapted from Example K of Cox and Snell (1981), taken in turn from John and Quenouille (1977). Table 5.1 shows the total weights of 24 six-week-old chicks. The treatments, twelve different methods of feeding, consisted of all combinations of three factors: level of protein at three levels, type of protein at two levels, and level of fish solubles, at two levels. The resulting $3 \times 2 \times 2$ factorial experiment was independently replicated in two different houses, which we treat as blocks in a randomized block experiment.

Table 5.2 shows mean responses cross-classified by factors. Tables of means are important both for a preliminary assessment of the data, and for summarizing the results. The average response on groundnut is 6763 g, and on soybean is 7012 g, which suggests that soybean is the more effective diet. However, the two-way table of type of protein by level is indicative of what will be called an interaction: the superiority of soybean appears to be reversed at the higher level of protein.

A plot for detecting interactions is often a helpful visual summary of the tables of means; see Figure 5.1, which is derived from Figure K.1 of Cox and Snell (1981). The interaction of type of

Table 5.1 *Total weights (g) of six-week-old chicks.*

Protein	Level of protein	Level of fish solubles	House I	House II	Mean
Groundnut	0	0	6559	6292	6425.5
		1	7075	6779	6927.0
	1	0	6564	6622	6593.0
		1	7528	6856	7192.0
	2	0	6738	6444	6591.0
		1	7333	6361	6847.0
Soybean	0	0	7094	7053	7073.5
		1	8005	7657	7831.0
	1	0	6943	6249	6596.0
		1	7359	7292	7325.5
	2	0	6748	6422	6585.0
		1	6764	6560	6662.0

protein with level of protein noted above shows in the lack of parallelism of the two lines corresponding to each type of protein. It is now necessary to check the strength of evidence for the effects just summarized. We develop definitions and methods for doing this in the next section.

5.3 Main effects and interactions

5.3.1 Assessing interaction

Consider two factors A and B at a, b levels, each combination replicated r times. Denote the response in the sth replicate at level i of A and j of B by Y_{ijs}. There are ab treatments, so the sum of squares for treatment has $ab - 1$ degrees of freedom and can be computed from the two-way table of means $\bar{Y}_{ij\cdot}$, averaging over s.

The first and primary analysis of the data consists of forming the two-way table of $\bar{Y}_{ij\cdot}$ and the associated one-way marginal means $\bar{Y}_{i\cdot\cdot}, \bar{Y}_{\cdot j\cdot}$, as we did in the example above. It is then important to determine if all the essential information is contained in comparison of the one-way marginal means, and if so, how the precision of

Table 5.2 *Two-way tables of mean weights (g).*

		Groundnut	Soybean	Mean
Level of	0	6676	7452	7064
protein	1	6893	6961	7927
	2	6719	6624	6671
Mean		6763	7012	6887

		G-nut	Soy	Level of protein			Mean
				0	1	2	
Level of	0	6537	6752	6750	6595	6588	6644
fish	1	6989	7273	7379	7259	6755	7131
Mean		6763	7012	7064	6927	6671	6887

associated contrasts is to be assessed. Alternatively, if more than one-way marginal means are important then appropriate more detailed interpretation will be needed.

We start the more formal analysis by assuming a linear model for Y_{ijs} that includes block effects whenever dictated by the design, and we define τ_{ij} to be the treatment effect for the treatment combination i, j. Using a dot here to indicate averaging over a subscript, we can write

$$\tau_{ij} = \tau_{..} + (\tau_{i.} - \tau_{..}) + (\tau_{.j} - \tau_{..}) + (\tau_{ij} - \tau_{i.} - \tau_{.j} + \tau_{..}), \quad (5.1)$$

and in slightly different notation

$$\tau_{ij} = \tau_{..} + \tau_i^A + \tau_j^B + \tau_{ij}^{AB}. \quad (5.2)$$

If the last term is zero for all i, j, then in the model the following statements are equivalent.

1. There is defined to be no interaction between A and B.
2. The effects of A and B are additive.
3. The difference between any two levels of A is the same at all levels of B.
4. The difference between any two levels of B is the same at all levels of A.

Of course, in the data there will virtually always be nonzero estimates of the above quantities. We define the sum of squares for

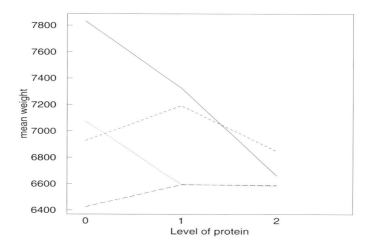

Figure 5.1 *Plot of mean weights (g) to show possible interaction. Soybean, Lev.f = 0 (———); Soybean, Lev.f = 1 (- - -); G-nut, Lev.f = 0 (······); G-nut, Lev.f = 1 (— — —)*

interaction via the marginal means corresponding to the last term in (5.1):

$$\sum_{i,j,s}(\bar{Y}_{ij.} - \bar{Y}_{i..} - \bar{Y}_{.j.} + \bar{Y}_{...})^2. \qquad (5.3)$$

Note that this is r times the corresponding sum over only i, j. One problem is to assess the significance of this, usually using an error term derived via the variation of effects between replicates, as in any randomized block experiment. If there were to be just one unit receiving each treatment combination, $r = 1$, then some other approach to estimating the variance is required.

An important issue of interpretation that arises repeatedly also in more complicated situations concerns the role of main effects in the presence of interaction.

Consider the interpretation of, say, $\tau_2^A - \tau_1^A$. When interaction is present the difference between levels 2 and 1 of A at level j of factor B, namely

$$\tau_{2j} - \tau_{1j} \qquad (5.4)$$

in the notation of (5.1), depends on j. If these individual differences have different signs then we say there is *qualitative* interaction. In the absence of qualitative interaction, the main effect of A, which is the average of the individual differences over j, retains some weak interpretation as indicating the general direction of the effect of A at all levels of B used in the experiment. However, generally in the presence of interaction, and especially qualitative interaction, the main effects do not provide a useful summary of the data and interpretation is primarily via the detailed pattern of individual effects.

In the definitions of main effects and interactions used above the parameters automatically satisfy the constraints

$$\Sigma_i \tau_i^A = \Sigma_j \tau_j^B = \Sigma_i \tau_{ij}^{AB} = \Sigma_j \tau_{ij}^{AB} = 0. \qquad (5.5)$$

The parameters are of three kinds and it is formally possible to produce submodels in which all the parameters of one, or even two, types are zero. The model with all τ_{ij}^{AB} zero, the model of main effects, has a clear interpretation and comparison with it is the basis of the test for interaction. The model with, say, all τ_j^B zero, i.e. with main effect of A and interaction terms in the model is, however, artificial and in almost all contexts totally implausible as a basis for interpretation. It would allow effects of B at individual levels of A but these effects would average *exactly* to zero over the levels of A that happened to be used in the experiment under analysis. Therefore, with rare exceptions, a hierarchical principle should be followed in which if an interaction term is included in a model so too should both the associated main effects. The principle extends when there are more than two factors.

A rare exception is when the averaging over the particular levels say of factor B used in the study has a direct physical significance.

Illustration. In an animal feeding trial suppose that A represents the diets under study and that B is not a treatment but an intrinsic factor, sex. Then interaction means that the difference between diets is not the same for males and females and qualitative interaction means that there are some reversals of effect, for example that diet 2 is better than diet 1 for males and inferior for females. Inspection only of the main effect of diets would conceal this. Suppose, however, that on the basis of the experiment a recommendation is to be made as to the choice of diet and this choice must for practical reasons be the same for male as for female animals. Suppose also that the target population has an equal mix of males and females.

Then regardless of interaction the main effect of diets should be the basis of choice. We stress that such a justification will rarely be available.

5.3.2 Higher order interaction

When there are more than two factors the above argument extends by induction. As an example, if there are three factors we could assess the three-way interaction $A \times B \times C$ by examining the two-way tables of $A \times B$ means at each level of C. If there is no three-factor interaction, then

$$\tau_{ijk} - \tau_{i.k} - \tau_{.jk} + \tau_{..k} \qquad (5.6)$$

is independent of k for all i, j, k and therefore is equivalent to

$$\tau_{ij.} - \tau_{i..} - \tau_{.j.} + \tau_{...}. \qquad (5.7)$$

We can use this argument to conclude that the three-way interaction, which is symmetric in i, j and k, should be defined in the model by

$$\tau_{ijk}^{ABC} = \tau_{ijk} - \tau_{ij.} - \tau_{i.k} - \tau_{.jk} + \tau_{i..} + \tau_{.j.} + \tau_{..k} - \tau_{...} \qquad (5.8)$$

and the corresponding sum of squares of the observations can be used to assess the significance of an interaction in data.

Note that these formal definitions apply also when one or more of the factors refer to properties of experimental units rather than to treatments. Testing the significance of interactions, especially higher order interactions, can be an important part of analysis, whereas for the main effects of treatment factors estimation is likely to be the primary focus of analysis.

5.3.3 Interpretation of interaction

Clearly lack of interaction greatly simplifies the conclusions, and in particular means that reporting the average response for each factor level is meaningful.

If there is clear evidence of interaction, then the following points will be relevant to the interpretation of the analysis. First, summary tables of means for factor A, say, averaged over factor B, or for $A \times B$ averaged over C will not be generally useful in the presence of interaction. As emphasized in the previous subsection the significance (or otherwise) of main effects is virtually always irrelevant in the presence of appreciable interaction.

Secondly, some particular types of interaction can be removed by transformation of the response. This indicates that a scale inappropriate for the interpretation of response may have been used. Note, however, that if the response variable analysed is physically additive, i.e. is extensive, transformation back to the original scale is likely to be needed for subject matter interpretation.

Thirdly, if there are many interactions involving a particular factor, separate analyses at the different levels of that factor may lead to the most incisive interpretation. This may especially be the case if the factor concerns intrinsic properties of the experimental units: it may be scientifically more relevant to do separate analyses for men and for women, for example.

Fourthly, if there are many interactions of very high order, there may be individual factor combinations showing anomalous response, in which case a factorial formulation may well not be appropriate.

Finally, if the levels of some or all of the factors are defined by quantitative variables we may postulate an underlying relationship $E\{Y(x_1, x_2)\} = \eta(x_1, x_2)$, in which a lack of interaction indicates $\eta(x_1, x_2) = \eta_1(x_1) + \eta_2(x_2)$, and appreciable interaction suggests a model such as

$$\eta(x_1, x_2) = \eta_1(x_1) + \eta_2(x_2) + \eta_{12}(x_1, x_2), \qquad (5.9)$$

where $\eta_{12}(x_1, x_2)$ is not additive in its arguments, for example depending on $x_1 x_2$. An important special case is where $\eta(x_1, x_2)$ is a quadratic function of its arguments. Response surface methods for problems of this sort will be considered separately in Section 6.6.

Our attitude to interaction depends considerably on context and indeed is often rather ambivalent. Interaction between two treatment factors, especially if it is not removable by a meaningful nonlinear transformation of response, is in one sense rather a nuisance in that it complicates simple description of effects and may lead to serious errors of interpretation in some of the more complex fractionated designs to be considered later. On the other hand such interactions may have important implications pointing to underlying mechanisms. Interactions between treatments and specific features of the experimental units, the latter in this context sometimes being called *effect modifiers*, may be central to interpretation and, in more applied contexts, to action. Of course interactions expected

Table 5.3 *Analysis of variance for a two factor experiment in a completely randomized design.*

Source	Sum of squares	Degrees of freedom
A	$\sum_{i,j,s}(\bar{Y}_{i..} - \bar{Y}_{...})^2$	$a-1$
B	$\sum_{i,j,s}(\bar{Y}_{.j.} - \bar{Y}_{...})^2$	$b-1$
$A \times B$	$\sum_{i,j,s}(\bar{Y}_{ij.} - \bar{Y}_{i..} - \bar{Y}_{.j.} + \bar{Y}_{...})^2$	$(a-1)(b-1)$
Residual	$\sum_{i,j,s}(\bar{Y}_{ijs} - \bar{Y}_{ij.})^2$	$ab(r-1)$

on *a priori* subject matter grounds deserve more attention than those found retrospectively.

5.3.4 Analysis of two factor experiments

Suppose we have two treatment factors, A and B, with a and b levels respectively, and we have r replications of a completely randomized design in these treatments. The associated linear model can be written as

$$Y_{ijs} = \mu + \tau_i^A + \tau_j^B + \tau_{ij}^{AB} + \epsilon_{ijs}. \quad (5.10)$$

The analysis centres on the interpretation of the table of treatment means, i.e. on the $\bar{Y}_{ij.}$ and calculation and inspection of this array is a crucial first step.

The analysis of variance table is constructed from the identity

$$\begin{aligned} Y_{ijs} &= \bar{Y}_{...} + (\bar{Y}_{i..} - \bar{Y}_{...}) + (\bar{Y}_{.j.} - \bar{Y}_{...}) \\ &+ (\bar{Y}_{ij.} - \bar{Y}_{i..} - \bar{Y}_{.j.} + \bar{Y}_{...}) + (Y_{ijs} - \bar{Y}_{ij.}). \end{aligned} \quad (5.11)$$

If the ab possible treatments have been randomized to the rab experimental units then the discussion of Chapter 4 justifies the use of the residual mean square, i.e. the variation between units within each treatment combination, as an estimate of error. If the experiment were arranged in randomized blocks or Latin squares or other similar design there would be a parallel analysis incorporating block, or row and column, or other relevant effects.

Thus in the case of a randomized block design in r blocks, we would have $r-1$ degrees of freedom for blocks and a residual with $(ab-1)(r-1)$ degrees of freedom used to estimate error, as in a simple randomized block design. The residual sum of squares,

EXAMPLE: CONTINUED

Table 5.4 *Analysis of variance for factorial experiment on the effect of diets on weights of chicks.*

Source	Sum of sq.	D.f.	Mean sq.
House	708297	1	708297
p-type	373751	1	373751
p-level	636283	2	318141
f-level	1421553	1	1421553
p-type× p-level	858158	2	429079
p-type× f-level	7176	1	7176
p-level× f-level	308888	2	154444
p-type× p-level× f-level	50128	2	25064
Residual	492640	11	44785

which is formally an interaction between treatments and blocks, can be partitioned into $A \times$ blocks, $B \times$ blocks and $A \times B \times$ blocks, giving separate error terms for the three components of the treatment effect. This would normally only be done if there were expected to be some departure from unit-treatment additivity likely to induce heterogeneity in the random variability. Alternatively the homogeneity of these three sums of squares provides a test of unit-treatment additivity, albeit one of low sensitivity.

In Section 6.5 we consider the interpretation when one or more of the factors represent nonspecific classification of the experimental units, for example referring to replication of an experiment over time, space, etc.

5.4 Example: continued

In constructing the analysis of variance table we treat House as a blocking factor, and assess the size of treatment effects relative to the interaction of treatments with House, the latter providing an appropriate estimate of variance for comparing treatment means, as discussed in Section 5.3. Table 5.4 shows the analysis of variance. As noted in Section 5.3 the residual sum of squares can be partitioned into components to check that no one effect is unusually large.

Since level of protein is a factor with three levels, it is possible

Table 5.5 *Decomposition of treatment sum of squares into linear and quadratic contrasts.*

Source	Sum of sq.	D.f.
House	708297	1
p-type	373751	1
p-level		
linear	617796	1
quadratic	18487	1
f-level	1421553	1
p-type× p-level		
linear	759510	1
quadratic	98640	1
p-type× f-level	7176	1
p-level× f-level		
linear	214370	1
quadratic	94520	1
p-type× p-level× f-level		
linear	47310	1
quadratic	2820	1
Residual	492640	11

to partition the two degrees of freedom associated with its main effects and its interactions into components corresponding to linear and quadratic contrasts, as outlined in Section 3.5. Table 5.5 shows this partition. From the three-way table of means included in Table 5.1 we see that the best treatment combination is a soybean diet at its lowest level, combined with the high level of fish solubles: the average weight gain on this diet is 7831 g, and the next best diet leads to an average weight gain of 7326 g. The estimated variance of the difference between two treatment means is $2(\tilde{\sigma}^2/2 + \tilde{\sigma}^2/2) = 211.6^2$ where $\tilde{\sigma}^2$ is the residual mean square in Table 5.4.

5.5 Two level factorial systems

5.5.1 General remarks

Experiments with large numbers of factors are often used as a screening device to assess quickly important main effects and interactions. For this it is common to set each factor at just two levels, aiming to keep the size of the experiment manageable. The levels of each factor are conventionally called low and high, or absent and present.

We denote the factors by $A, B, \ldots$ and a general treatment combination by $a^i b^j \ldots$, where $i, j, \ldots$ take the value zero when the corresponding factor is at its low level and one when the corresponding factor is at its high level. For example in a 2^5 design, the treatment combination bde has factors A and C at their low level, and B, D and E at their high level. The treatment combination of all factors at their low level is (1).

We denote the treatment means in the population, i.e. the expected responses under each treatment combination, by $\mu_{(1)}, \mu_a$, and so on. The observed response for each treatment combination is denoted by $Y_{(1)}, Y_a$, and so on. These latter will be averages over replicates if there is more than one observation on each treatment.

The simplest case is a 2^2 experiment, with factors A and B, and four treatment combinations $(1), a, b$, and ab. There are thus four identifiable parameters, the general mean, two main effects and an interaction. In line with the previous notation we denote these by μ, τ^A, τ^B and τ^{AB}. The population treatment means $\mu_{(1)}, \mu_a, \mu_b, \mu_{ab}$ are simple linear combinations of these parameters:

$$\tau^A = (\mu_{ab} + \mu_a - \mu_b - \mu_{(1)})/4,$$
$$\tau^B = (\mu_{ab} - \mu_a + \mu_b - \mu_{(1)})/4,$$
$$\tau^{AB} = (\mu_{ab} - \mu_a - \mu_b + \mu_{(1)})/4.$$

The corresponding least squares estimate of, for example, τ^A, under the summation constraints, is

$$\hat{\tau}^A = (Y_{ab} + Y_a - Y_b - Y_{(1)})/4$$
$$= (\bar{Y}_{2..} - \bar{Y}_{1..})/2 = \bar{Y}_{2..} - \bar{Y}_{...} \quad (5.12)$$

where, for example, $\bar{Y}_{2..}$ is the mean over all replicates and over both levels of factor B of the observations taken at the higher level of A. Similarly

$$\hat{\tau}^{AB} = (Y_{ab} - Y_a - Y_b + Y_{(1)})/4 = (\bar{Y}_{11.} - \bar{Y}_{21.} - \bar{Y}_{12.} + \bar{Y}_{22.})/4, \quad (5.13)$$

where $\bar{Y}_{11.}$ is the mean of the r observations with A and B at their lower levels.

The A contrast, also called the A *main effect*, is estimated by the difference between the average response among units receiving high A and the average response among units receiving low A, and is equal to $2\hat{\tau}^A$ as defined above. In the notation of (5.2) $\hat{\tau}^A = \hat{\tau}_2^A = -\hat{\tau}_1^A$, and the estimated A effect is defined to be $\hat{\tau}_2^A - \hat{\tau}_1^A$. The interaction is estimated via the difference between $Y_{ab} - Y_b$ and $Y_a - Y_{(1)}$, i.e. the difference between the effect of A at the high level of B and the effect of A at the low level of B.

Thus the estimates of the effects are specified by three orthogonal linear contrasts in the response totals. This leads directly to an analysis of variance table of the form shown in Table 5.6.

By defining

$$I = (1/4)(\mu_{(1)} + \mu_a + \mu_b + \mu_{ab}) \qquad (5.14)$$

we can write

$$\begin{bmatrix} I \\ A \\ B \\ AB \end{bmatrix} = \begin{matrix} (1/4) \\ (1/2) \\ (1/2) \\ (1/2) \end{matrix} \begin{bmatrix} 1 & 1 & 1 & 1 \\ -1 & 1 & -1 & 1 \\ -1 & -1 & 1 & 1 \\ 1 & -1 & -1 & 1 \end{bmatrix} \begin{bmatrix} \mu_{(1)} \\ \mu_a \\ \mu_b \\ \mu_{ab} \end{bmatrix} \qquad (5.15)$$

and this pattern is readily generalized to k greater than 2; for example

$$\begin{bmatrix} 8I \\ 4A \\ 4B \\ 4AB \\ 4C \\ 4AC \\ 4BC \\ 4ABC \end{bmatrix} = \begin{bmatrix} 1 & 1 & 1 & 1 & 1 & 1 & 1 & 1 \\ -1 & 1 & -1 & 1 & -1 & 1 & -1 & 1 \\ -1 & -1 & 1 & 1 & -1 & -1 & 1 & 1 \\ 1 & -1 & -1 & 1 & 1 & -1 & -1 & 1 \\ -1 & -1 & -1 & -1 & 1 & 1 & 1 & 1 \\ 1 & -1 & 1 & -1 & -1 & 1 & -1 & 1 \\ 1 & 1 & -1 & -1 & -1 & -1 & 1 & 1 \\ -1 & 1 & 1 & -1 & 1 & -1 & -1 & 1 \end{bmatrix} \begin{bmatrix} \mu_{(1)} \\ \mu_a \\ \mu_b \\ \mu_{ab} \\ \mu_c \\ \mu_{ac} \\ \mu_{bc} \\ \mu_{abc} \end{bmatrix}. \qquad (5.16)$$

Note that the effect of AB, say, is the contrast of $a^i b^j c^k$ for which $i + j = 0 \bmod 2$, with those for which $i + j = 1 \bmod 2$. Also the product of the coefficients for C and ABC gives the coefficients for AB, etc. All the contrasts are orthogonal.

The matrix in (5.16) is constructed row by row, the first row consisting of all 1's. The rows for A and B have entries $-1, +1$ in the order determined by that in the set of population treatment means: in (5.16) they are written in *standard order* to make construction of the matrix straightforward. The row for AB is the product of those for A and B, and so on. Matrices for up to a 2^6 design can be quickly tabulated in a table of signs.

Table 5.6 *Analysis of variance for r replicates of a 2^2 factorial.*

Source	Sum sq.	D.f.
Factor A	SS_A	1
Factor B	SS_B	1
Interaction	SS_{AB}	1
Residual		$4(r-1)$
Total		$4r-1$

5.5.2 General definitions

The matrix approach outlined above becomes increasingly cumbersome as the number of factors increases. It is convenient for describing the general 2^k factorial to use some group theory: Appendix B provides the basic definitions. The treatment combinations in a 2^k factorial form a *prime power commutative* group; see Section B.2.2. The set of contrasts also forms a group, dual to the treatment group.

In the 2^3 factorial the *treatment group* is

$$\{(1), a, b, ab, c, ac, bc, abc\}$$

and the *contrast group* is

$$\{I, A, B, AB, C, AC, BC, ABC\}.$$

As in (5.16) above, each contrast is the difference of the population means for two sets of treatments, and the two sets of treatments are determined by an element of the contrast group. For example the element A partitions the treatments into the sets $\{(1), b, c, bc\}$ and $\{a, ab, ac, abc\}$, and the A effect is thus defined to be $(\mu_a + \mu_{ab} + \mu_{ac} + \mu_{abc} - \mu_{(1)} - \mu_b - \mu_c - \mu_{bc})/4$.

In a 2^k factorial we define a contrast group $\{I, A, B, AB, \ldots\}$ consisting of symbols $A^\alpha B^\beta C^\gamma \cdots$, where $\alpha, \beta, \gamma, \ldots$ take values 0 and 1. An arbitrary nonidentity element $A^\alpha B^\beta \cdots$ of the contrast group divides the treatments into two sets, with $a^i b^j c^k \cdots$ in one set or the other according as

$$\alpha i + \beta j + \gamma k + \cdots = 0 \bmod 2, \quad (5.17)$$
$$\alpha i + \beta j + \gamma k + \cdots = 1 \bmod 2. \quad (5.18)$$

Then
$$A^\alpha B^\beta C^\gamma \cdots = \frac{1}{2^{k-1}}\{\text{sum of } \mu'\text{s in set containing } a^\alpha b^\beta c^\gamma \cdots$$
$$- \text{sum of } \mu'\text{s in other set}\}, \quad (5.19)$$
$$I = \frac{1}{2^k}\{\text{sum of all } \mu'\text{s}\}. \quad (5.20)$$

The two sets of treatments defined by any contrast form a subgroup and its coset; see Section B.2.2.

More generally, we can divide the treatments into 2^l subsets using a contrast subgroup of order 2^l. Let S_C be a subgroup of order 2^l of the contrast group defined by l generators

$$\begin{aligned} G_1 &= A^{\alpha_1} B^{\beta_1} \cdots \\ G_2 &= A^{\alpha_2} B^{\beta_2} \cdots \\ &\vdots \\ G_l &= A^{\alpha_l} B^{\beta_l} \cdots. \end{aligned} \quad (5.21)$$

Divide the treatments group into 2^l subsets containing

(i) all symbols with (even, even, ... , even) number of letters in common with $G_1, \ldots, G_l$;

(ii) all symbols with (odd, even, ... , even) number of letters in common with $G_1, \ldots, G_l$

$\vdots$

(2^l) all symbols with (odd, odd, ... , odd) number of letters in common with $G_1, \ldots, G_l$.

Then (i) is a subgroup of order 2^{k-l} of the treatments group, and all sets (ii) ... (2^l) contain 2^{k-l} elements and are cosets of (i). In particular, therefore, there are the same number of treatments in each of these sets.

For example, in a 2^4 design, the contrasts ABC, BCD (or the contrast subgroup $\{I, ABC, BCD, AD\}$) divide the treatments into the four sets

(i) $\{(1), bc, abd, acd\}$
(ii) $\{a, abc, bd, cd\}$
(iii) $\{d, bcd, ab, ac\}$
(iv) $\{ad, abcd, b, c\}$.

The treatment subgroup in (i) is dual to the contrast subgroup.

TWO LEVEL FACTORIAL SYSTEMS 115

Any two contrasts are orthogonal, in the sense that the defining contrasts divide the treatments into four equally sized sets, a subgroup and three cosets.

5.5.3 Estimation of contrasts

In a departure from our usual practice, we use the same notation for the population contrast and for its estimate. Consider a design in which each of the 2^k treatments occurs r times, the design being arranged in completely randomized form, or in randomized blocks with 2^k units per block, or in $2^k \times 2^k$ Latin squares. The least squares estimates of the population contrasts are simply obtained by replacing population means by sample means: for example,

$$A^\alpha B^\beta C^\gamma \cdots = \frac{1}{2^{k-1}r}\{\text{sum of } y's \text{ in set containing } a^\alpha b^\beta c^\gamma \cdots$$
$$- \text{ sum of } y's \text{ in other set}\}, \quad (5.22)$$

$$I = \frac{1}{2^k r}\{\text{sum of all } y's\}. \quad (5.23)$$

Each contrast is estimated by the difference of two means each of $r2^{k-1} = (1/2)n$ observations, which has variance $2\sigma^2/(r2^{k-1}) = 4\sigma^2/n$. The analysis of variance table, for example for the randomized blocks design, is given in Table 5.7. The single degree of freedom sums of squares are equal to $r2^{k-2}$ times the square of the corresponding estimated effect, a special case of the formula for a linear contrast given in Section 3.5. If meaningful, the residual sum of squares can be partitioned into sets of $r - 1$ degrees of freedom. A table of estimated effects and their standard errors will usually be a more useful summary than the analysis of variance table.

Typically for moderately large values of k the experiment will not be replicated, so there is no residual sum of squares to provide a direct estimate of the variance. A common technique is to pool the estimated effects of the higher order interactions, the assumption being that these interactions are likely to be negligible, in which case each of their contrasts has mean zero and variance $4\sigma^2/n$. If we pool l such estimated effects, we have l degrees of freedom to estimate σ^2. For example, in a 2^5 experiment there are five main effects and 10 two factor interactions, leaving 16 residual degrees of freedom if all the third and higher order interactions are pooled.

A useful graphical aid is a normal probability plot of the estimated effects. The estimated effects are ordered from smallest

Table 5.7 *Analysis of variance for a 2^k design.*

Source	Degrees of freedom
Blocks	$r-1$
Treatments	$2^k - 1 \begin{cases} 1 \\ \vdots \\ 1 \end{cases}$
Residual	$(r-1)(2^k - 1)$

to largest, and the ith effect in this list of size $2^k - 1$ is plotted against the expected value of the ith largest of $2^k - 1$ order statistics from the standard normal distribution. Such plots typically have a number of nearly zero effect estimates falling on a straight line, and a small number of highly significant effects which readily stand out. Since all effects have the same estimated variance, this is an easy way to identify important main effects and interactions, and to suggest which effects to pool for the estimation of σ^2. Sometimes further plots may be made in which either all main effects are omitted or all manifestly significant contrasts omitted. The expected value of the ith of n order statistics from the standard normal can be approximated by $\Phi^{-1}\{i/(n+1)\}$, where $\Phi(\cdot)$ is the cumulative distribution function of the standard normal distribution. A variation on this graphical aid is the *half normal plot*, which ranks the estimated effects according to their absolute values, which are plotted against the corresponding expected value of the absolute value of a standard normal variate. The full normal plot is to be preferred if, for example, the factor levels are defined in such a way that the signs of the estimated main effects have a reasonably coherent interpretation, for example that positive effects are *a priori* more likely than negative effects.

5.6 Fractional factorials

In some situations quite sharply focused research questions are formulated involving a small number of key factors. Other factors may be involved either for technical reasons, or to explore interactions,

FRACTIONAL FACTORIALS   117

but the contrasts of main concern are clear. In other applications of a more exploratory nature, there may be a large number of factors of potential interest and the working assumption is often that only main effects and a small number of low order interactions are important. Other possibilities are that only a small group of factors and their interactions influence response or that response may be the same except when all factors are simultaneously at their high levels.

Illustration. Modern techniques allow the modification of single genes to find the gene or genes determining a particular feature in experimental animals. For some features it is likely that only a small number of genes are involved.

We turn now to methods particularly suited for the second situation mentioned above, namely when main effects and low order interactions are of primary concern.

A complete factorial experiment with a large number of factors requires a very large number of observations, and it is of interest to investigate what can be estimated from only part of the full factorial experiment. For example, a 2^7 factorial requires 128 experimental units, and from these responses there are to be estimated 7 main effects and 21 two factor interactions, leaving 99 degrees of freedom to estimate error and/or higher order interactions. It seems feasible that quite good estimates of main effects and two factor interactions could often be obtained from a much smaller experiment.

As a simple example, suppose in a 2^3 experiment we obtain observations only from treatments (1), ab, ac and bc. The linear combination $(y_{ab} + y_{ac} - y_{bc} - y_{(1)})/2$ provides an estimate of the A contrast, as it compares all observations at the high level of A with those at the low level. However, this linear combination is also the estimate that would be obtained for the interaction BC, using the argument outlined in the previous section. The main effect of A is said to be *aliased* with that of BC. Similarly the main effect of B is aliased with AC and that of C aliased with AB. The experiment that consists in obtaining observations only on the four treatments (1), ab, ac and bc is called a *half-fraction* or *half-replicate* of a 2^3 factorial.

The general discussion in Section 5.5 is directly useful for defining a 2^{-l} fraction of a 2^k factorial. These designs are called 2^{k-l} fractional factorials. Consider first a 2^{k-1} fractional factorial. As we saw in Section 5.5.2, any element of the contrast group partitions the treatments into two sets. A half-fraction of the 2^k factorial consists of the experiment taking observations on one of these two sets. The contrast that is used to define the sets cannot be estimated from the experiment, but every other contrast can be, as all the constrasts are orthogonal. For example, in a 2^5 factorial we might use the contrast $ABCDE$ to define the two sets. The set of treatments $a^i b^j c^k d^l e^m$ for which $i+j+k+l+m = 0$ mod 2 forms the first half fraction.

More generally, any subgroup of order 2^l of the contrast group, defined by l generators, divides the treatments into a subgroup and its cosets. A 2^{k-l} fractional factorial design takes observations on just one of these sets of treatments, say the subgroup, set (i). Now consider the estimation of an arbitrary contrast $A^\alpha B^\beta \ldots$. This compares treatments for which

$$\alpha i + \beta j + \ldots = 0 \bmod 2 \qquad (5.24)$$

with those for which

$$\alpha i + \beta j + \ldots = 1 \bmod 2. \qquad (5.25)$$

However, by construction of the treatment subgroup all treatments satisfy $\alpha_r i + \beta_r j + \ldots = 0$ mod 2 for $r = 1, \ldots 2^l - 1$, see (5.21), so that we are equally comparing

$$(\alpha + \alpha_r)i + (\beta + \beta_r)j + \ldots = 0 \qquad (5.26)$$

with

$$(\alpha + \alpha_r)i + (\beta + \beta_r)j + \ldots = 1. \qquad (5.27)$$

Thus any estimated contrast has $2^l - 1$ alternative interpretations, i.e. aliases, obtained by multiplying the contrast into the elements of the *alias subgroup*. The general theory is best understood by working through an example in detail: see Exercise 5.2.

In general we aim to choose the alias subgroup so that so far as possible main effects and two factor interactions are aliased with three factor and higher order interactions. Such a design is called a design of Resolution V; designs in which two factor interactions are aliased with each other are Resolution IV, and designs in which two factor interactions are aliased with main effects are Resolution III. The resolution of a fractional factorial is equal to the length of the shortest member of the alias subgroup.

For example, suppose that we wanted a 1/4 replicate of a 2^6

FRACTIONAL FACTORIALS   119

factorial, i.e. a 2^{6-2} design investigating six factors in 16 observations. At first sight it might be tempting to take five factor interactions to define the aliasing subgroup, for example taking $ABCDE$, $BCDEF$ as generators leading to the contrast subgroup

$$\{I, ABCDE, BCDEF, AF\}, \tag{5.28}$$

clearly a poor choice for nearly all purposes because the main effects of A and F are aliased. A better choice is the Resolution IV design with contrast subgroup

$$\{I, ABCD, CDEF, ABEF\} \tag{5.29}$$

leaving each main effect aliased with two three-factor interactions. Some two factor interactions are aliased in triples, e.g. AB, CD and EF, and others in pairs, e.g. AC and BD, and occasionally some use could be made of this distinction in naming the treatments. To find the 16 treatment combinations forming the design we have to find four independent generators of the appropriate subgroup and form the full set of treatments by repeated multiplication. The choice of particular generators is arbitrary but might be ab, cd, ef, ace yielding

$$\begin{array}{llllllll} (1) & ab & cd & abcd & ef & abef & cdef & abcdef \\ ace & bce & ade & bde & acf & bcf & adf & bdf \end{array} \tag{5.30}$$

A coset of these could be used instead.

Note that if, after completing this 1/4 replicate, it were decided that another 1/4 replicate is needed, replication of the same set of treatments would not usually be the most suitable procedure. If, for example it were of special interest to clarify the status of the interaction AB, it would be sensible in effect to reduce the aliasing subgroup to

$$\{I, CDEF\} \tag{5.31}$$

by forming a coset by multiplication by, for example, acd, which is not in the above subgroup but which is even with respect to $CDEF$.

There are in general rich possibilities for the formation of series of experiments, clarifying at each stage ambiguities in the earlier results and perhaps removing uninteresting-seeming factors and adding new ones.

5.7 Example

Blot et al. (1993) report a large nutritional intervention trial in Linxian county in China. The goal was to investigate the role of dietary supplementation with specific vitamins and minerals on the incidence of and mortality from esophageal and stomach cancers, a leading cause of mortality in Linxian county. There were nine specific nutrients of interest, but a 2^9 factorial experiment was not considered feasible. The supplements were instead administered in combination, and each of four factors was identified by a particular set of nutrients, as displayed in Table 5.8.

The trial recruited nearly 30 000 residents, who were randomly assigned to receive one of eight vitamin/mineral supplement combinations within blocks defined by commune, sex and age. The treatment set formed a one-half fraction of the full 2^4 design with the contrast $ABCD$ defining the fraction. Table 5.9 shows the data, the number of cancer deaths and person-years of observation in each of the eight treatment groups. Estimates of the main effects and the two factor interactions are presented in Table 5.10. The two factor interactions are all aliased in pairs.

Table 5.8 *Treatment factors: combinations of micronutrients. From Blot et al. (1993).*

Factor	Micronutrients	Dose per day
A	Retinol	5000 IU
	Zinc	22.5 mg
B	Riboflavin	3.2 mg
	Niacin	40 mg
C	Vitamin C	120 mg
	Molybdenum	30 μg
D	Beta carotene	15 mg
	Selenium	50 μg
	Vitamin E	30 mg

We estimate the treatment effects using a log-linear model for the rates of cancer deaths in the eight groups. If we regard the counts as being approximately Poisson distributed, the variance of the log of a single response is approximately $1/\mu$, where μ is the Poisson

EXAMPLE

Table 5.9 *Cancer mortality in the Linxian study. From Blot et al. (1993).*

Treatment	Person-years of observation, n_c	Number of cancer deaths, d_c	Deaths from all causes
(1)	18626	107	280
ab	18736	94	265
ac	13701	121	296
bc	18686	101	268
ad	18745	81	250
bd	18729	103	263
cd	18758	90	249
$abcd$	18792	95	256

Table 5.10 *Estimated effects based on analysis of* $\log(d_c/n_c)$.

A	B	C	D	AB, CD
-0.036	-0.005	0.053	-0.140	-0.043

AC, BD	AD, BC
0.152	-0.058

mean (Exercise 8.3), so the average of these across the eight groups is estimated by $\frac{1}{8}(1/107 + \ldots + 1/95) = 0.010$. Since each contrast is the difference between averages of four totals, the standard error of the estimated effects is approximately 0.072. From this we see that the main effect of D is substantial, although the interpretation of this is somewhat confounded by the large increase in mortality rate associated with the interaction $AC = BD$. This is consistent with the conclusion reached in the analysis of Blot et al. (1993), that dietary supplementation with beta carotene, selenium and vitamin E is potentially effective in reducing the mortality from stomach cancers. There is a similar effect on total mortality, which is analysed in Appendix C.

To some extent this analysis sets aside one of the general principles of Chapters 2 and 3. By treating the random variation as having a Poisson distribution we are in effect treating individual subjects as the experimental units rather than the groups of subjects which are the basis of the randomization. It is thus assumed that so far as the contrasts of treatments are concerned the blocking has essentially accounted for all the overdispersion relative to the Poisson distribution that is likely to be present. The more careful analysis of Blot et al. (1993), which used the more detailed data in which the randomization group is the basis of the analysis, essentially confirms that.

5.8 Bibliographic notes

The importance of a factorial approach to the design of experiments was a key element in Fisher's (1926, 1935) systematic approach to the subject. Many of the crucial details were developed by Yates (1935, 1937). A review of the statistical aspects of interaction is given in Cox (1984a); see also Cox and Snell (1981; Section 4.13). For discussion of qualitative interaction, see Azzalini and Cox (1984), Gail and Simon (1985) and Ciminera et al. (1993).

Factorial experiments are quite widely used in many fields. For a review of the fairly limited number of clinical trials that are factorial, see Piantadosi (1997, Chapter 15). The systematic exploration of factorial designs in an industrial context is described by Box, Hunter and Hunter (1978); see also the Bibliographic notes for Chapter 6. Daniel (1959) introduced the graphical method of the half-normal plot; see Olguin and Fearn (1997) for the calculation of guard rails as an aid to interpretation.

Fractional replication was first discussed by Finney (1945b) and, independently, for designs primarily concerned with main effects, by Plackett and Burman (1945). For an introductory account of the mathematical connection between fractional replication and coding theory, see Hill (1986) and also the Bibliographic notes for Appendix B.

A formal mathematical definition of the term *factor* is provided in McCullagh (2000) in relation to category theory. This provides a mathematical interpretation to the notion that main effects are not normally meaningful in the presence of interaction. McCullagh also uses the formal definition of a factor to emphasize that associated models should preserve their form under extension and contraction of the set of levels. This is particularly relevant when some of the factors are homologous, i.e. have levels with identical meanings.

5.9 Further results and exercises

1. For a single replicate of the 2^2 system, write the observations as a column vector in the standard order $1, a, b, ab$. Form a new 4×1 vector by adding successive pairs and then subtracting successive pairs, i.e. to give $Y_1 + Y_a, Y_b + Y_{ab}, Y_a - Y_1, Y_{ab} - Y_b$. Repeat this operation on the new vector and check that there results, except for constant multipliers, estimates in standard order of the contrast group.

   Show by induction that for the 2^k system k repetitions of the above procedure yield the set of estimated contrasts.

   Observe that the central operation is repeated multiplication by the 2×2 matrix

   $$M = \begin{pmatrix} 1 & 1 \\ 1 & -1 \end{pmatrix} \quad (5.32)$$

   and that the kth Kronecker product of this matrix with itself generates the matrix defining the full set of contrasts.

   Show further that by working with a matrix proportional to M^{-1} we may generate the original observations starting from the set of contrasts and suggest how this could be used to smooth a set of observations in the light of an assumption that certain contrasts are null.

   The algorithm was given by Yates (1937) after whom it is commonly named. An extension covering three level factors is due to Box and reported in the book edited by Davies (1956). The elegant connection to Kronecker products and the fast Fourier transform was discussed by Good (1958).

2. Construct a 1/4 fraction of a 2^5 factorial using the generators $G_1 = ABCD$ and $G_2 = CDE$. Write out the sets of aliased effects.

3. Using the construction outlined in Section 5.5.2 in a 2^4 factorial, verify that any contrast does define two sets of treatments, with 2^3 treatments in each set, and that any pair of contrasts divides the treatments into four sets each of 2^2 treatments.

4. In the notation of Section 5.5.2, verify that the 2^l subsets of the treatment group constructed there are equally determined by the conditions:

   (i) $\quad \alpha_1 i + \beta_1 j + \ldots = 0, \quad \alpha_2 i + \beta_2 j + \ldots = 0, \quad \ldots,$

$$
\begin{aligned}
&\text{(ii)} \quad &&\alpha_l i + \beta_l j + \ldots = 0, \\
& &&\alpha_1 i + \beta_1 j + \ldots = 1, \quad \alpha_2 i + \beta_2 j + \ldots = 0, \quad \ldots, \\
& &&\alpha_l i + \beta_l j + \ldots = 0, \\
& &&\quad \vdots \\
&(2^l) \quad &&\alpha_1 i + \beta_1 j + \ldots = 1, \quad \alpha_2 i + \beta_2 j + \ldots = 1, \quad \ldots, \\
& &&\alpha_l i + \beta_l j + \ldots = 1.
\end{aligned}
$$

5. Table 5.11 shows the design and responses for four replicates of a 1/4 fraction of a 2^6 factorial design. The generators used to determine the set of treatments were $ABCD$ and $BCEF$. Describe the alias structure of the design and discuss its advantages and disadvantages. The factors represent the amounts of various minor ingredients added to flour during milling, and the response variable is the average volume in ml/g of three loaves of bread baked from dough using the various flours (Tuck, Lewis and Cottrell, 1993). Table 5.12 gives the estimates of the main effects and estimable two factor interactions. The standard error of the estimates can be obtained by pooling small effects or via the treatment-block interaction, treating day as a blocking factor. The details are outlined in Appendix C.

6. Factorial experiments are normally preferable to those in which successive treatment combinations are defined by changing only one factor at a time, as they permit estimation of interactions as well as of main effects. However, there may be cases, for example when it is very difficult to vary factor levels, where one-factor-at-a-time designs are needed. Show that for a 2^3 factorial, the design which has the sequence of treatments $(1), a, ab, abc, bc, c, (1)$ permits estimation of the three main effects and of the three interaction sums $AB + AC$, $-AB + BC$ and $AC + BC$. This design also has the property that the main effects are not confounded by any linear drift in the process over the sequence of the seven observations. Extend the discussion to the 2^4 experiment (Daniel, 1994).

7. In a fractional factorial with a largish number of factors, there may be several designs of the same resolution. One means of choosing between them rests on the combinatorial concept of *minimum aberration* (Fries and Hunter, 1980). For example a fractional factorial of resolution three has no two factor interactions aliased with main effects, but they may be aliased with

Table 5.11 *Exercise 5.6: Volumes of bread (ml/g) from Tuck, Lewis and Cottrell (1993).*

Factor levels; factors are coded ingredient amounts						Average specific volume for the following days:			
A	B	C	D	E	F	1	2	3	4
−1	−1	−1	−1	−1	−1	519	446	337	415
−1	−1	−1	−1	1	1	503	468	343	418
−1	−1	1	1	−1	1	567	471	355	424
−1	−1	1	1	−1	−1	552	489	361	425
−1	1	−1	1	−1	1	534	466	356	431
−1	1	−1	1	1	−1	549	461	354	427
−1	1	1	−1	−1	−1	560	480	345	437
−1	1	1	−1	1	1	535	477	363	418
1	−1	−1	1	−1	−1	558	483	376	418
1	−1	−1	1	1	1	551	472	349	426
1	−1	1	−1	−1	1	576	487	358	434
1	−1	1	−1	1	−1	569	494	357	444
1	1	−1	−1	−1	1	562	474	358	404
1	1	−1	−1	1	−1	569	494	348	400
1	1	1	1	−1	−1	568	478	367	463
1	1	1	1	1	1	551	500	373	462

Table 5.12 *Estimates of contrasts for data in Table 5.11.*

A	B	C	D	E	F	AB	
13.66	3.72	14.72	7.03	−0.16	−2.41	−2.53	
AC	BC	AE	BE	CE	DE	ABE	ACE
0.22	−2.84	−0.1	0.03	0.16	−0.66	3.16	2.53

each other. In this setting the design of minimum aberration equalizes as far as possible the number of two factor interactions in each alias set. See Dey and Mukerjee (1999, Chapter 8) and Cheng and Mukerjee (1998) for a more detailed discussion. Another method for choosing among fractional factorial designs is to minimize (or conceivably maximize) the number of level changes required during the execution of the experiment. See

Cheng, Martin and Tang (1998) for a mathematical discussion. Mesenbrink et al. (1994) present an interesting case study in which it was very expensive to change factor levels.

CHAPTER 6

Factorial designs: further topics

6.1 General remarks

In the previous chapter we discussed the key ideas involved in factorial experiments and in particular the notions of interaction and of the possibility of extracting useful information from fractions of the full factorial system. We begin the present more specialized chapter with a discussion of confounding, mathematically closely connected with fractional replication but conceptually quite different. We continue with various more specialized topics related to factorial designs, including factors at more than two levels, orthogonal arrays, split unit designs, and response surface methods.

6.2 Confounding in 2^k designs

6.2.1 Simple confounding

Factorial and fractional factorial experiments may require a large number of experimental units, and it may thus be advisable to use one or more of the methods described in Chapters 3 and 4 for controlling haphazard variation. For example, it may be feasible to try only eight treatment combinations in a given day, or four treatment combinations on a given batch of raw material. The treatment sets defined in Section 5.5.2 may then be used to arrange the 2^k experimental units in blocks of size 2^{k-p} in such a way that block differences can be eliminated without losing information about specified contrasts, usually main effects and low order interactions.

For example, in a 2^3 experiment to be run in two blocks, we can use the ABC effect to define the blocks, by simply putting into one block the treatment subgroup obtained by the contrast subgroup $\{I, ABC\}$, and into the second block its coset:

Block 1:	(1)	ab	ac	bc
Block 2:	a	b	c	abc

Note that the second block can be obtained by multiplication mod 2 of any one element of that block with those in the first block. The ABC effect is now confounded with blocks, i.e. it is not possible to estimate it separately from the block effect. The analysis of variance table has one degree of freedom for blocks, and six degrees of freedom for the remaining effects A, B, C, AB, AC, BC. The whole experiment could be replicated r times.

Experiments with larger numbers of factors can be divided into a larger number of blocks by identifying the subgroup and cosets of the treatment group associated with particular contrast subgroups. For example, in a 2^5 experiment the two contrasts ABC, CDE form the contrast subgroup $\{I, ABC, CDE, ABDE\}$, and this divides the treatment group into the following sets:

Block 1:	(1)	ab	acd	bcd	ace	bce	de	$abde$
Block 2:	a	b	cd	$abcd$	ce	$abce$	ade	bde
Block 3:	c	abc	ad	bd	ae	be	cde	$abcde$
Block 4:	ac	bc	d	abd	e	abe	$acde$	$bcde$

following the discussion of Section 5.5.2. The defining contrasts ABC, CDE, and their product mod 2, namely $ABDE$, are confounded with blocks. If there were prior information to indicate that particular interactions are of less interest than others, they would, of course, be chosen as the ones to be confounded.

With larger experiments and larger blocks the general discussion of Section 5.5.2 applies directly. The block that receives treatment (1) is called the *principal block*. In the analysis of a 2^k experiment run in 2^p blocks, we have $2^p - 1$ degrees of freedom for blocks, and $2^k - 2^p$ estimated effects that are not confounded with blocks. Each of the unconfounded effects is estimated in the usual way, as a difference between two equal sized sets of treatments, divided by $r2^{k-1}$ if there are r replicates, and estimated with variance $\sigma^2/(r2^{k-2})$. A summary is given in a table of estimated effects and their standard errors, or for some purposes in an analysis of variance table. If, as would often be the case, there is just one replicate of the experiment ($r = 1$), the error can be estimated by pooling higher order unconfounded interactions, as discussed in Section 5.5.3.

If there are several replicates of the blocked experiment, and the same contrasts are confounded in all replicates, they are said to be

CONFOUNDING IN 2^k DESIGNS

totally confounded. Using the formulae from Section 5.5.3, the unconfounded contrasts are estimated with standard error $2\sigma_m/\sqrt{n}$ where n is the total number of units and σ_m^2 is the variance among responses in a single block of m units, where here $m = 2^{k-p}$. If the experiment were not blocked, the corresponding standard error would be $2\sigma_n/\sqrt{n}$, where σ_n^2 is the variance among all n units, and would often be appreciably larger than σ_m^2.

6.2.2 Partial confounding

If we can replicate the experiment, then it may be fruitful to confound different contrasts with blocks in different replicates, in which case we can recover an estimate of the confounded interactions, although with reduced precision. For example, if we have four replicates of a 2^3 design run in two blocks of size 4, we could confound ABC in the first replicate, AB in the second, AC in the third, and BC in the fourth, giving:

Replicate I:	Block 1:	(1)	ab	ac	bc
	Block 2:	a	b	c	abc
Replicate II:	Block 1:	(1)	ab	c	abc
	Block 2:	a	b	bc	ac
Replicate III:	Block 1:	(1)	b	ac	abc
	Block 2:	a	c	bc	ab
Replicate IV:	Block 1:	(1)	a	bc	abc
	Block 2:	b	c	ab	ac

Estimates of a contrast and its standard error are formed from the replicates in which that contrast is not confounded. In the above example we have three replicates from which to estimate each of the four interactions, and four replicates from which to estimate the main effects. Thus if σ_m^2 denotes the error variance corresponding to blocks of size m, all contrasts are estimated with higher precision after confounding provided $\sigma_4^2/\sigma_8^2 < 3/4$.

A further fairly direct development is to combine fractional replication with confounding. This is illustrated in Section 6.3 below.

6.2.3 Double confounding

In special cases it may be possible to construct orthogonal confounding patterns using different sets of contrasts, and then to

Table 6.1 *An example of a doubly confounded design.*

(1)	abcd	bce	ade	acf	bdf	abef	cdef
abd	c	acde	be	bcdf	af	def	abcef
abce	de	a	bcd	bef	acdef	cf	abdf
cde	abe	bd	ac	adef	bcef	abcdf	f
bcf	adf	ef	abcdef	ab	cd	ace	bde
acdf	bf	abd	cef	d	abc	bcde	ae
aef	bcdef	abcf	df	ce	abde	b	acd
bdef	acef	cdf	abf	abcde	e	ad	bc

associate the sets of treatments so defined with two (or more) different blocking factors, for example the rows and columns of a Latin square style design. The following example illustrates the main ideas.

In a 2^6 experiment suppose we choose to confound ACE, ADF, and BDE with blocks. This divides the treatments into blocks of size 8, and the interactions $CDEF$, $ABCD$, $ABEF$ and BCF are also confounded with blocks. The alternative choice ABF, ADE, and BCD also determines blocks of size 8, with a distinct set of confounded interactions (BDE, $ACDF$, $ABCE$ and CEF). Thus we can use both sets of generators to set out the treatments in a $2^3 \times 2^3$ square. The design before randomization is shown in Table 6.1. The principal block for the first confounding pattern gives the first row, the principal block for the second confounding pattern gives the first column, and the remaining treatment combinations are determined by multiplication (mod 2) of these two sets of treatments, achieving coset structure both by rows and by columns.

The form of the analysis is summarized in Table 6.2, where the last three rows would usually be pooled to give an estimate of error with 22 degrees of freedom.

Fractional factorial designs may also be laid out in blocks, in which case the effects defining the blocks and all their aliases are confounded with blocks.

OTHER FACTORIAL SYSTEMS

Table 6.2 *Degrees of freedom for the doubly confounded Latin square design in Table 6.1.*

Rows	7
Columns	7
Main effects	6
2-factor interactions	15
3-factor interactions	20 − 8=12
4-factor interactions	15 − 6 = 9
5-factor interactions	6
6-factor interaction	1

6.3 Other factorial systems

6.3.1 General remarks

It is often necessary to consider factorial designs with factors at more than two levels. Setting a factor at three levels allows, when the levels are quantitative, estimation of slope and curvature, and thus, in particular, a check of linearity of response. A factor with four levels can formally be regarded as the product of two factors at two levels each, and the design and analysis outlined in Chapter 5 can be adapted fairly directly.

For example, a 3^2 design has factors A and B at each of three levels, say 0, 1 and 2. The nine treatment combinations are (1), a, a^2, b, b^2, ab, a^2b, ab^2 and a^2b^2. The main effect for A has two degrees of freedom and is estimated from two contrasts, preferably but not necessarily orthogonal, between the total response at the three levels of A. If the factor is quantitative it is natural to use the linear and quadratic contrasts with coefficients $(-1, 0, 1)$ and $(1, -2, 1)$ respectively (cf. Section 3.5). The $A \times B$ interaction has four degrees of freedom, which might be decomposed into single degrees of freedom using the direct product of the same pair of contrast coefficients. The four components of interaction are denoted $A_L B_L$, $A_L B_Q$, $A_Q B_L$, $A_Q B_Q$, in an obvious notation. If the levels of the two factors were indexed by x_1 and x_2 respectively, then these four effects are coefficients of the products $x_1 x_2$, $x_1 x_2^2$, $x_1^2 x_2$, and $(x_1^2 - 1)(x_2^2 - 1)$. The first effect is essentially the interaction component of the quadratic term in the response, to which the cubic and quartic effects are to be compared.

Table 6.3 *Two orthogonal Latin squares used to partition the $A \times B$ interaction.*

Q	R	S		Q	R	S
R	S	Q		S	Q	R
S	Q	R		R	S	Q

A different partition of the interaction term is suggested by considering the two orthogonal 3×3 Latin squares shown in Table 6.3. If we associate the levels of A and B with respectively the rows and the columns of the squares, the letters essentially identify the treatment combinations $a^i b^j$. Each square gives two degrees of freedom for (P, Q, R), so that the two factor interaction has been partitioned into two components, written formally as AB and AB^2. These components have no direct statistical interpretation, but can be used to define a confounding scheme if it is necessary to carry out the experiment in three blocks of size three, or to define a 3^{2-1} fraction.

6.3.2 Factors at a prime number of levels

Consider experiments in which all factors occur at a prime number p of levels, where $p = 3$ is the most important case. The mathematical theory for $p = 2$ generalizes very neatly, although it is not too satisfactory statistically.

The treatment combinations $a^i b^j \ldots$, where i and j run from 0 to $p - 1$, form a group $\mathcal{G}_p(a, b, \ldots)$ with the convention $a^p b^p \ldots = 1$; see Appendix B. If we form a table of totals of observations as indicated in Table 6.4, we define the main effect of A, denoted by the symbols $A, \ldots, A^{p-1}$ to be the $p-1$ degrees of freedom involved in the contrasts among the p totals. This set of degrees of freedom is defined by contrasting the p sets of $a^i b^j \ldots$ for $i = 0, 1, \ldots, p-1$.

To develop the general case we assume familiarity with the Galois field of order p, $\mathrm{GF}(p)$, as sketched in Appendix B.3. In general let $\alpha, \beta, \gamma, \ldots \in \mathrm{GF}(p)$ and define

$$\phi = \alpha i + \beta j + \cdots . \tag{6.1}$$

This sorts the treatments into sets defined by $\phi = 0, 1, \ldots, p - 1$. The sets can be shown to be equal in size. Hence ϕ determines a

OTHER FACTORIAL SYSTEMS

Table 6.4 *Estimation of the main effect of A.*

	(Sum over $b, c, \ldots$)	
a^0	a^1	a^{p-1}
$Y_{0\ldots}$	$Y_{1\ldots}$	$Y_{p-1\ldots}$

contrast with $p-1$ degrees of freedom. Clearly $c\phi$ determines the same contrast. We denote it by $A^\alpha B^\beta \cdots$ or equally $A^{c\alpha} B^{c\beta} \cdots$, where $c = 1, \ldots, p-1$. By convention we arrange that the first non-zero coefficient is a one. For example, with $p = 5$, $B^3 C^2, BC^4, B^4 C$ and $B^2 C^3$ all represent the same contrast. The conventional form is BC^4.

We now suppose in (6.1) that $\alpha \neq 0$. Consider another contrast defined by $\phi' = \alpha' i + \beta' j + \ldots$, and suppose $\alpha' \neq 0$. Among all treatments satisfying $\phi = c$, and for fixed $j, k, \ldots$, we have $i = (c - \beta j - \gamma k - \ldots)/\alpha$ and then eliminating i from ϕ' gives

$$\phi' = \frac{\alpha'}{\alpha} c + (\beta' - \frac{\alpha'}{\alpha}\beta)j + \ldots \quad (6.2)$$

with not all the coefficients zero. As $j, k, \ldots$ run through all values, with ϕ fixed, so does ϕ'. Hence the contrasts defined by ϕ' are orthogonal to those defined by ϕ.

We have the following special cases

1. For the main effect of A, $\phi = i$.
2. In the table of AB totals there are $p^2 - 1$ degrees of freedom. The main effects account for $2(p-1)$. The remaining $(p-1)^2$ form the interaction $A \times B$. They are the contrasts $AB, AB^2, \ldots, AB^{p-1}$ each with $p-1$ degrees of freedom.
3. Similarly $ABC, ABC^2, \ldots, AB^{p-1}C^{p-1}$ are $(p-1)^2$ sets of $(p-1)$ degrees of freedom each, forming the $A \times B \times C$ interaction with $(p-1)^3$ degrees of freedom.

The limitation of this approach is that the subdivision of, say, the $A \times B$ interaction into separate sets of degrees of freedom usually has no statistical interpretation. For example, if the factor levels were determined by equal spacing of a quantitative factor, this subdivision would not correspond to a partition by orthogonal polynomials, which is more natural.

In the 3^2 experiment discussed above, the main effect A compares $a^i b^j$ for $i = 0, 1, 2 \mod 3$, the interaction AB compares $a^i b^j$ for $i + j = 0, 1, 2 \mod (3)$, and the interaction AB^2 compares $a^i b^j$ for $i + 2j = 0, 1, 2 \mod (3)$. We can set this out in two orthogonal 3×3 Latin squares as was done above in Table 6.3.

In a 3^3 experiment the two factor interactions such as $B \times C$ are split into pairs of degrees of freedom as above. Now consider the $A \times B \times C$ interaction. This is split into:

$$\begin{aligned}
ABC &: i + j + k = 0, 1, 2 \mod 3 \\
ABC^2 &: i + j + 2k = 0, 1, 2 \mod 3 \\
AB^2C &: i + 2j + k = 0, 1, 2 \mod 3 \\
AB^2C^2 &: i + 2j + 2k = 0, 1, 2 \mod 3
\end{aligned}$$

We may consider the ABC term for example, as determined from a Latin cube, laid out as follows:

$$\begin{array}{ccc} Q & R & S \\ R & S & Q \\ S & Q & R \end{array} \qquad \begin{array}{ccc} R & S & Q \\ S & Q & R \\ Q & R & S \end{array} \qquad \begin{array}{ccc} S & Q & R \\ Q & R & S \\ R & S & Q \end{array}$$

where in the first layer Q corresponds to the treatment combination with $i + j + k = 0$, R with $i + j + k = 1$, and S with $i + j + k = 2$. There are three further Latin cubes, orthogonal to the above cube, corresponding to the three remaining components of interaction ABC^2, AB^2C, and AB^2C^2. In general with r letters we have $(p-1)^{r-1}$ r-dimensional orthogonal $p \times p$ Latin hypercubes.

Each contrast divides the treatments into three equal sets and can therefore be a basis for confounding. Thus ABC divides the 3^3 experiment into three blocks of nine, and with four replicates we can confound in turn ABC^2, AB^2C, and AB^2C^2. Similarly taking $\{I, ABC\}$ as defining an alias subgroup, the nine treatments Q above form a $\frac{1}{3}$ replicate with

$$\begin{aligned}
I &= ABC = A^2 B^2 C^2 \\
A &= A^2 BC = B^2 C^2 (= BC) \\
B &= AB^2 C = A^2 C^2 (= AC) \\
C &= ABC^2 = A^2 B^2 (= AB) \\
AB^2 &= A^2 C (= AC^2) = BC^2, \quad \text{etc.}
\end{aligned}$$

OTHER FACTORIAL SYSTEMS

Factors at p^m levels can be regarded as the product of m factors at p levels or dealt with directly by $\mathrm{GF}(p^m)$; see Appendix B. The case of four levels is sketched in Exercise 6.4.

For example, the 3^5 experiment has a $\frac{1}{3}$ replicate such that aliases of all main effects and two factor interactions are higher order interactions; for example we can take as the alias subgroup $\{I, ABCDE, A^2B^2C^2D^2E^2\}$. This can be confounded into three blocks of 27 units each using ABC^2 as the effect to be confounded with blocks, giving

$$ABC^2 = A^2B^2DE = CD^2E^2,$$
$$A^2B^2C = C^2DE = ABD^2E^2.$$

The contents of the first block must satisfy $i + j + k + l + m = 0 \bmod (3)$ and $i + j + 2k = 0 \bmod (3)$, which gives three generators. The treatments in this block are

(1) de^2 d^2e ab^2 a^2b ab^2de^2 ab^2d^2e a^2bde^2 a^2bd^2e
acd acd^2e^2 acd^2e bcd b^2cd $a^2b^2cd^2e^2$ a^2b^2ce bcd^2e^2
bce $a^2c^2d^2$ $a^2c^2e^2$ a^2c^2de abc^2d^2 $b^2c^2d^2$ $b^2c^2e^2$ b^2c^2de
abc^2e^2 abc^2de

The second and third blocks are found by multiplying by treatments that satisfy $i + j + k + l + m = 0$, but not $i + j + 2k = 0$. Thus ad^2 and a^2d^2 will achieve this. The analysis of variance is set out in Table 6.5.

Table 6.5 *Degrees of freedom for the estimable effects in a confounded 1/3 replicate of a 3^5 design.*

Source	D.f.
Blocks	2
Main effects	10
Two factor interactions = three factor interactions	$\left(\frac{5\times 4}{1\times 2}\right) \times 2 = 20$
Two factor interactions = four factor interactions	20
Three factor interactions ($\neq$ two factor interactions)	$\frac{5\times 4\times 3}{1\times 2\times 3} \times 3 \times 2 \times \frac{1}{2} - 2 = 28$
Total	80

The Latin square has entered the discussion at various points. The design was introduced in Chapter 4 as a design for a single set of treatments with the experimental units cross-classified by rows and by columns. No special assumptions were involved in its analysis. By contrast if we have three treatment factors all with the same number, k, of levels we can regard a $k \times k$ Latin square as a one-kth replicate of the k^3 system in which main effects can be estimated separately, assuming there to be no interactions between the treatments. Yet another role of a $k \times k$ Latin square is as a one-kth replicate of a k^2 system in k randomized blocks. These should be thought of as three quite distinct designs with a common combinatorial base.

6.3.3 Orthogonal arrays

We consider now the structure of factorial or fractional factorial designs from a slightly different point of view. We define for a factorial experiment an *orthogonal array*, which is simply a matrix with runs or experimental units indexing the rows, and factors indexing the columns. The elements of the array indicate the level of the factors for each run. For example, the orthogonal array associated with a single replicate of a 2^3 factorial may be written out as

$$\begin{array}{rrr} -1 & -1 & -1 \\ 1 & -1 & -1 \\ -1 & 1 & -1 \\ 1 & 1 & -1 \\ -1 & -1 & 1 \\ 1 & -1 & 1 \\ -1 & 1 & 1 \\ 1 & 1 & 1 \end{array} \qquad (6.3)$$

We could as well use the symbols $(0, 1)$ as elements of the array, or ("high", "low"), etc. The structure of the design is such that the columns are mutually orthogonal, and in any pair of columns each possible treatment combination occurs the same number of times. The columns of the array (6.3) are the three rows in the matrix of contrast coefficients (5.16) corresponding to the three main effects of factors A, B, and C.

The full array of contrast coefficients is obtained by pairwise

OTHER FACTORIAL SYSTEMS

multiplication of columns of (6.3):

$$
\begin{array}{cccccccc}
1 & -1 & -1 & 1 & -1 & 1 & 1 & -1 \\
1 & 1 & -1 & -1 & -1 & -1 & 1 & 1 \\
1 & -1 & 1 & -1 & -1 & 1 & -1 & 1 \\
1 & 1 & 1 & 1 & -1 & -1 & -1 & -1 \\
1 & -1 & -1 & 1 & 1 & -1 & -1 & 1 \\
1 & 1 & -1 & -1 & 1 & 1 & -1 & -1 \\
1 & -1 & 1 & -1 & 1 & -1 & 1 & -1 \\
1 & 1 & 1 & 1 & 1 & 1 & 1 & 1 \\
\hline
 & A & B & C & D & E & F & G
\end{array}
\qquad (6.4)
$$

As indicated by the letters across the bottom, we can associate a main effect with each column except the first, in which case (6.4) defines a 2^{7-4} factorial with, for example, $C = AB = EF$, etc. Array (6.4) is an 8×8 Hadamard matrix; see Appendix B. Each row indexes one run or one experimental unit. For example, the first run has factors A, B, D, G at their low level and the others at their high level. The main effects of factors A up to G are independently estimable by the indicated contrasts in the eight observations: for example the main effect of E is estimated by $(Y_1 - Y_2 + Y_3 - Y_4 - Y_5 + Y_6 + Y_7 - Y_8)/4$. This design is called *saturated* for main effects; once the main effects have been estimated there are no degrees of freedom remaining to estimate interactions or error.

An orthogonal array of size $n \times n-1$ with two symbols in each column specifies a design saturated for main effects. The designs with symbols ± 1 are called Plackett-Burman designs and Hadamard matrices defining them have been shown to exist for all multiples of four up to 424; see Appendix B.

More generally, an $n \times k$ array with m_i symbols in the ith column is an *orthogonal array* of strength r if all possible combinations of symbols appear equally often in any r columns. The symbols correspond to levels of a factor. The array in (6.3) has 2 levels in each column, and has strength 2, as each of $(-1,-1)$, $(-1,+1)$, $(+1,-1)$, $(+1,+1)$ appears the same number of times in every set of two columns. An orthogonal array with all m_i equal is called symmetric. The strength of the array is a generalization of the notion of resolution of a fractional factorial, and determines the number of independent estimable effects.

Table 6.6 gives an asymmetric orthogonal array of strength 2 with $m_1 = 3$, $m_2 = m_3 = m_4 = 2$. Each level of each factor occurs

Table 6.6 *An asymmetric orthogonal array.*

−1	−1	−1	−1	−1
−1	−1	1	−1	1
−1	1	−1	1	1
−1	1	1	1	−1
0	−1	−1	1	1
0	−1	1	1	−1
0	1	−1	−1	1
0	1	1	−1	−1
1	−1	−1	1	−1
1	−1	1	−1	1
1	1	−1	−1	−1
1	1	1	1	1
A	B	C	D	E

the same number of times with each level of the remaining factors. Thus, for example, linear and quadratic effects of A and B can be estimated, as well as the linear effects used in specifying the design.

There is a large literature on the existence and construction of orthogonal arrays; see the Bibliographic notes. Methods of construction include ones based on orthogonal Latin squares, on difference matrices, and on finite projective geometries. Orthogonal arrays of strength 2 are often associated with Taguchi methods, and are widely used in industrial experimentation; see Section 6.7.

6.3.4 Supersaturated systems

In an experiment with n experimental units and k two-level factors it may if $n = k + 1$ be possible to find a design in which all main effects can be estimated separately, for example by a fractional factorial design with main effects aliased only with interactions. Indeed this is possible, for example when $k = 2^m - 1$ using the orthogonal arrays described in the previous subsection. Such designs are saturated with main effects.

OTHER FACTORIAL SYSTEMS

Table 6.7 *Supersaturated design for 16 factors in 12 trials.*

+	+	+	+	+	+	+	+	+	+	+	−	−	−	−	−
+	−	+	+	+	−	−	−	+	−	−	−	−	−	−	−
−	+	+	+	−	−	−	+	−	−	+	+	+	−	+	+
+	+	+	−	−	−	+	−	−	+	−	−	+	+	+	+
+	+	−	−	−	+	−	−	+	−	+	+	+	−	+	−
+	−	−	−	+	−	−	+	−	+	+	+	+	+	−	+
−	−	−	+	−	−	+	−	+	+	+	+	−	+	+	+
−	−	+	−	−	+	−	+	+	+	−	+	−	+	−	+
−	+	−	−	+	−	+	+	+	−	−	+	+	+	+	−
+	−	−	+	−	+	+	+	−	−	−	−	+	+	−	−
−	−	+	−	+	+	+	−	−	−	+	−	−	−	+	+
−	+	−	+	+	+	−	−	−	+	−	−	−	−	−	−

Suppose now that $n < k + 1$, i.e. that there are fewer experimental units than parameters in a main effects model. A design for such situations is called *supersaturated*. For example we might want to study 16 factors in 12 units. Clearly all main effects cannot be separately estimated in such situations. If, however, to take an extreme case, it could plausibly be supposed that at most one factor has a nonzero effect, it will be possible with suitable design to isolate that factor. If we specify the design by a $n \times k$ matrix of 1's and -1's it is reasonable to make the columns as nearly mutually orthogonal as possible. Such designs may be found by computer search or by building on the theory of fractional replication.

These designs are not merely sensitive to the presence of interactions aliased with main effects but more seriously still if more than a rather small number of effects are present very misleading conclusions may be drawn.

Table 6.7 shows a design for 16 factors in 12 trials. It was formed by adding to a main effect design for 11 factors five additional columns obtained by computer search. First the maximum scalar product of two columns was minimized. Then, within all designs with the same minimum, the number of pairs of columns with that value was minimized.

140   FACTORIAL DESIGNS: FURTHER TOPICS

While especially in preliminary industrial investigations it is entirely possible that the number of factors of potential interest is more than the number of experimental units available for an initial experiment, it is questionable whether the use of supersaturated designs is ever the most sensible approach. Two alternatives are abstinence, cutting down the number of factors in the initial study, and the use of judicious factor amalgamation. For the latter suppose that two factors A and B are such that their upper and lower levels can be defined in such a way that if either has an effect it is likely to be that the main effect is positive. We can then define a new two-level quasi-factor (AB) with levels $(1), (ab)$ in the usual notation. If a positive effect is found for (AB) then it is established that at least one of A and B has an effect. In this way the main effects of factors of particular interest and which are not amalgamated are estimated free of main effect aliasing, whereas other main effects have a clear aliasing structure. Without the assumption about the direction of any effect there is the possibility of effect cancellation. Thus in examining 16 factors in 12 trials we would aim to amalgamate 10 factors in pairs and to investigate the remaining 6 factors singly in a design for 11 new factors in 12 trials.

6.4 Split plot designs

6.4.1 General remarks

Formally a *split plot*, or *split unit*, experiment is a factorial experiment in which a main effect is confounded with blocks. There is, however, a difference of emphasis from the previous discussion of confounding. Instead of regarding the confounded main effects as lost, we now suppose there is sufficient replication for them to be estimated, although with lower, and maybe much lower, precision. In this setting blocks are called *whole units*, and what were previously called units are now called *subunits*. The replicates of the design applied to the whole units and subunits typically correspond to our usual notion of blocks, such as days, operators, and so on.

As an example suppose in a factorial experiment with two factors A and B, where A has four levels and B has three, we assign the following treatments to each of four blocks:

$$(1) \quad b \quad b^2 \qquad a \quad b \quad ab^2 \qquad a^2 \quad a^2b \quad a^2b^2 \qquad a^3 \quad a^3b \quad a^3b^2$$

SPLIT PLOT DESIGNS

in an obvious notation. The main effect of A is clearly confounded with blocks. Equivalently, we may assign the level of A at random to blocks or whole units, each of which consists of three subunits. The levels of B are assigned at random to the units in each block.

Now consider an experiment with, say kr whole units arranged in r blocks of size k. Let each whole unit be divided into s equal subunits. Let there be two sets of treatments (the simplest case being when there are two factors) and suppose that:

1. whole-unit treatments, $A_1, \ldots, A_k$, say, are applied at random in randomized block form to the whole units;

2. subunit treatments, $B_1, \ldots, B_s$, are applied at random to the subunits, each subunit treatment occurring once in each whole unit.

An example of one block with $k = 4$ and $s = 5$ is:

A_1	A_2	A_3	A_4
B_4	B_2	B_1	B_5
B_3	B_1	B_2	B_4
B_5	B_5	B_3	B_3
B_1	B_3	B_4	B_2
B_2	B_4	B_5	B_1

All the units in the same column receive the same level of A. There will be a similar arrangement, independently randomized, in each of the r blocks.

We can first do an analysis of the whole unit treatments represented schematically by:

Source	D.f.
Blocks	$r - 1$
Whole unit treatment A	$k - 1$
Error (a)	$(k-1)(r-1)$
Between whole units	$kr - 1$

The error is determined by the variation between whole units within blocks and the analysis is that of a randomized block design. We can now analyse the subunit observations as:

Between whole units	$kr - 1$
Subunit treatment B	$s - 1$
$A \times B$	$(s-1)(k-1)$
Error (b)	$k(r-1)(s-1)$
Total	$krs - 1$

The error (b) measures the variation between subunits within whole units. Usually this error is appreciably smaller than the whole unit error (a).

There are two reasons for using split unit designs. One is practical convenience, particularly in industrial experiments on two (or more) stage processes, where the first stage represents the whole unit treatments carried out on large batches, which are then split into smaller sections for the second stage of processing. This is the situation in the example discussed in Section 6.4.2. The second is to obtain higher precision for estimating B and the interaction $A \times B$ at the cost of lower precision for estimating A. As an example of this A might represent varieties of wheat, and B fertilisers: if the focus is on the fertilisers, two or more very different varieties may be included primarily to examine the $A \times B$ interaction thereby, hopefully, obtaining some basis for extending the conclusions about B to other varieties.

There are many variants of the split unit idea, such as the use of split-split unit experiments, subunits arranged in Latin squares, and so on. When we have a number of factors at two levels each we can apply the theory of Chapter 5 to develop more complicated forms of split unit design.

6.4.2 Examples

We first consider two examples of factorial split-unit designs. For the first example, let there be four two-level factors, and let it be required to treat one, A, as a whole unit treatment, the main effects of B, C, and D being required among the subunit treatments. Suppose that each replicate is to consist of four whole units, each containing four subunits. Take as the confounding subgroup $\{I, A, BCD, ABCD\}$. Then the design is, before randomization,

SPLIT PLOT DESIGNS

(1)	bc	cd	bd
a	abc	acd	abd
ab	ac	abcd	ad
b	c	bcd	d

As a second example, suppose we have five factors and that it is required to have $\frac{1}{2}$ replicates consisting of four whole units each of four subunits, with factor A having its main effect in the whole unit part. In the language of 2^k factorials we want a $\frac{1}{2}$ replicate of a 2^5 in 2^2 blocks of 2^2 units each with A confounded. The alias subgroup is $\{I, ABCDE\}$ with confounding subgroups

$$A = BCDE, \quad BC = ADE, \quad ABC = DE. \tag{6.5}$$

This leaves two two factor interactions in the whole unit part and we choose them to be those of least potential interest. The design is

(1)	bc	de	bcde
ab	ac	abde	acde
cd	bd	ce	be
ae	abce	ad	abcd

The analysis of variance table has the form outlined in Table 6.8. A prior estimate of variance will be necessary for this design.

Table 6.8 *Analysis of variance for the 5 factor example.*

	Source	D.f.
Between whole plots	A	1
	BC	1
	DE	1
Main effects	B,C,D,E	4
Two factor interactions		8 (= 10 − 2)
		15

Our third example illustrates the analysis of a split unit experiment, and is adapted from Montgomery (1997, Section 12.4). The

Table 6.9 *Tensile strength of paper. From Montgomery (1997).*

		Day 1			Day 2			Day 3		
Prep. method		1	2	3	1	2	3	1	2	3
	1	30	34	29	28	31	31	31	35	32
Temp	2	35	41	26	32	36	30	37	40	34
	3	37	38	33	40	42	32	41	39	39
	4	36	42	36	41	40	40	40	44	45

experiment investigated two factors, pulp preparation method and temperature, on the tensile strength of paper. Temperature was to be set at four levels, and there were three preparation methods. It was desired to run three replicates, but only 12 runs could be made per day. One replicate was run on each of the three days, and replicates (or days) is the blocking factor.

On each day, three batches of pulp were prepared by the three different methods; thus the level of this factor determines the whole unit treatment. Each of the three batches was subdivided into four equal parts, and processed at a different temperature, which is thus the subunit treatment. The data are given in Table 6.9.

The analysis of variance table is given in Table 6.10. If F-tests are of interest, the appropriate test for the main effect of preparation method is $64.20/9.07$, referred to an $F_{2,4}$ distribution, whereas for the main effect of temperature and the temperature $\times$ preparation interaction the relevant denominator mean square is 3.97. Similarly, the standard error of the estimated preparation effect is larger than that for the temperature and temperature $\times$ preparation effects. Estimates and their standard errors are summarized in Table 6.11.

6.5 Nonspecific factors

We have already considered the incorporation of block effects into the analysis of a factorial experiment set out in randomized blocks. This follows the arguments based on randomization theory and developed in Chapters 3 and 4. Formally a simple randomized block experiment with a single set of treatments can be regarded as one replicate of a factorial experiment with one treatment factor and

Table 6.10 *Analysis of variance table for split unit example.*

Source	Sum of sq.	D.f.	Mean sq.
Blocks	77.55	2	38.78
Prep. method	128.39	2	64.20
Blk × Prep.(error (a))	36.28	4	9.07
Temp	434.08	3	144.69
Prep× Temp	75.17	6	12.53
Error (b)	71.49	18	3.97

Table 6.11 *Means and estimated standard errors for split unit experiment.*

		Prep 1	Prep 2	Prep 3	Mean	
Temp	1	29.67	33.33	30.67	31.22	
	2	34.67	39.00	30.00	34.56	Standard error
	3	39.33	39.67	34.67	37.89	for difference 0.94
	4	39.00	42.00	40.33	40.44	
Mean		35.67	38.50	33.92	36.03	
		Standard error for difference 1.23				

one factor, namely blocks, referring to the experimental units. We call such a factor *nonspecific* because it will in general not be determined by a single aspect, such as sex, of the experimental units. In view of the assumption of unit-treatment additivity we may use the formal interaction, previously called residual, as a base for estimating the effective error variance. From another point of view we are imposing a linear model with an assumed zero interaction between treatments and blocks and using the associated residual mean square to estimate variance. In the absence of an external estimate of variance there is little effective alternative, unless some especially meaningful components of interaction can be identified and removed from the error estimate. But so long as the initial assumption of unit-treatment additivity is reasonable we need no special further assumption.

Now suppose that an experiment, possibly a factorial experiment, is repeated in a number of centres, for example a number of laboratories or farms or over a number of time points some appreciable way apart. The assumption of unit-treatment additivity across a wide range of conditions is now less appealing and considerable care in interpretation is needed.

Illustrations. Some agricultural field trials are intended as a basis for practical recommendations to a broad target population. There is then a strong case for replication over a number of farms and over time. The latter gives a spread of meteorological conditions and the former aims to cover soil types, farm management practices and so on. Clinical trials, especially of relatively rare conditions, often need replication across centres, possibly in different countries, both to achieve some broad representation of conditions, but also in order to accrue the number of patients needed to achieve reasonable precision.

To see the issues involved in fairly simple form suppose that we start with an experiment with just one factor A with r replicates of each treatment, i.e. in fact a simple nonfactorial experiment. Now suppose that this design is repeated independently at k centres; these may be different places, laboratories or times, for example. Formally this is now a two factor experiment with replication. We assume the effect of factors A and B on the expected response are of the form

$$\tau_{ij} = \tau_i^A + \tau_j^B + \tau_{ij}^{AB}, \tag{6.6}$$

using the notation of Section 5.3, and we compute the analysis of variance table by the obvious extension to the decomposition of the observations for the randomized block design Y_{ijs} used in Section 3.4:

$$\begin{aligned} Y_{ijs} &= \bar{Y}_{...} + (\bar{Y}_{i..} - \bar{Y}_{...}) - (\bar{Y}_{.j.} - \bar{Y}_{...}) \\ &\quad + (\bar{Y}_{ij.} - \bar{Y}_{i..} - \bar{Y}_{.j.} + \bar{Y}_{...}) + (Y_{ijs} - \bar{Y}_{ij.}). \end{aligned} \tag{6.7}$$

We can compute the expected mean squares from first principles under the summation restrictions $\Sigma \tau_i^A = 0$, $\Sigma \tau_j^B = 0$, $\Sigma_i \tau_{ij}^{AB} = 0$, and $\Sigma_j \tau_{ij}^{AB} = 0$. Then, for example, $E(\text{MS}_{AB})$ is equal to

$$E\{r\Sigma_{ij}(\bar{Y}_{ij.} - \bar{Y}_{i..} - \bar{Y}_{.j.} + \bar{Y}_{...})^2\}/\{(v-1)(k-1)\}$$
$$= rE\Sigma_{ij}\{\tau_{ij}^{AB} + (\bar{\epsilon}_{ij.} - \bar{\epsilon}_{i..} - \bar{\epsilon}_{.j.} + \bar{\epsilon}_{...})\}^2/\{(v-1)(k-1)\}$$
$$= r\Sigma_{ij}(\tau_{ij}^{AB})^2 + \{r\Sigma_{ij}E(\bar{\epsilon}_{ij.} - \bar{\epsilon}_{i..} - \bar{\epsilon}_{.j.} + \bar{\epsilon}_{...})^2\}/\{(v-1)(k-1)\}.$$

NONSPECIFIC FACTORS

The last expectation is that of a quadratic form in $\bar{\epsilon}_{ij\cdot}$ of rank $(v-1)(k-1)$ and hence equal to $\sigma^2(v-1)(k-1)/r$.

The analysis of variance table associated with this system has the form outlined in Table 6.12. From this we see that the design permits testing of $A \times B$ against the residual within centres. If unit-treatment additivity held across the entire investigation the interaction mean square and the residual mean square would both be estimates of error and would be of similar size; indeed if such unit-treatment additivity were specified the two terms would be pooled. In many contexts, however, it would be expected *a priori* and found empirically that the interaction mean square is greater than the mean square within centres, establishing that the treatment effects are not identical in the different centres.

If such an interaction is found, it should be given a rational interpretation if possible, either qualitatively or, for example, by finding an explicit property of the centres whose introduction into a formal model would account for the variation in treatment effect. In the absence of such an explanation there is little quantitative alternative to regarding the interaction as a haphazard effect represented by a random variable in an assumed linear model. Note that we would not do this if centres represented a specific property of the experimental material, and certainly not if centres had been a treatment factor.

A modification to the usual main effect and interaction model is

Table 6.12 *Analysis of variance for a replicated two factor experiment.*

Source	D.f.	Expected Mean squares
A, Trtms	$v-1$	$\sigma^2 + rk\Sigma(\tau_i^A)^2/(v-1)$
B, centres	$k-1$	$\sigma^2 + rv\Sigma(\tau_j^B)^2/(k-1)$
$A \times B$	$(v-1)(k-1)$	$\sigma^2 + r\Sigma(\tau_{ij}^{AB})^2/\{(v-1)(k-1)\}$
Within centres	$vk(r-1)$	σ^2

now essential. We write instead of (6.6)

$$\tau_{ij} = \tau^A_{\pi i} + \tau^B_j + \eta^{AB}_{ij}, \qquad (6.8)$$

where η^{AB}_{ij} are assumed to be random variables with zero mean, uncorrelated and with constant variance σ^2_{AB}, representing the haphazard variation in treatment effect from centre to centre. Note the crucial point that it would hardly ever make sense to force these haphazard effects to sum to zero over the particular centres used. There are, moreover, strong homogeneity assumptions embedded in this specification: in addition to assuming constant variance we are also excluding the possibility that there may be some contrasts that are null across all centres, and at the same time some large treatment effects that are quite different in different centres. If that were the case, the null effects would in fact be estimated with much higher precision than the non-null treatment effects and the treatment times centres interaction effect would need to be subdivided.

In (6.8) $\tau^A_{\pi 2} - \tau^A_{\pi 1}$ specifies the contrast of two levels averaged out not only over the differences between the experimental units employed but also over the distribution of the η^{AB}_{ij}, i.e. over a hypothetical ensemble π of repetitions of the centres.

A commonly employed, but in some contexts rather unfortunate, terminology is to call centres a random factor and to add the usually irrelevant assumption that the τ^B_j also are random variables. The objection to that terminology is that farms, laboratories, hospitals, etc. are rarely a random sample in any meaningful sense and, more particularly, if this factor represents time it is not often meaningful to regard time variation as totally random and free of trends, serial correlations, etc. On the other hand the approximation that the way treatment effects vary across centres is represented by uncorrelated random variables is weaker and more plausible.

The table of expected mean squares for model (6.8) is given in Table 6.13. The central result is that when interest focuses on treatment effects averaged over the additional random variation the appropriate error term is the mean square for interaction of treatments with centres. The arguments against study of the treatment main effect averaged over the particular centres in the study have already been rehearsed; if that was required we would, however, revert to the original specification and use the typically smaller

Table 6.13 *Analysis of variance for a two factor experiment with a random effect.*

Source	D.f.	Expected mean squares
A	$v-1$	$\sigma^2 + r\sigma_{AB}^2 + rk\Sigma(\tau_{\pi i}^A)^2/(v-1)$
B	$k-1$	$\sigma^2 + rv\Sigma(\tau_j^B)^2/(k-1)$
$A \times B$	$(v-1)(k-1)$	$\sigma^2 + r\sigma_{AB}^2$
residual	$vk(r-1)$	σ^2

mean square within centres to estimate the error variance associated with the estimation of the parameters $\tau_{\pi i}^A$.

6.6 Designs for quantitative factors

6.6.1 General remarks

When there is a single factor whose levels are defined by a quantitative variable, x, there is always the possibility of using a transformation of x to simplify interpretation, for example by achieving effective linearity of the dependence of the response on x or on powers of x. If a special type of nonlinear response is indicated, for example by theoretical considerations, then fitting by maximum likelihood, often equivalent to nonlinear least squares, will be needed and the methods of nonlinear design sketched in Section 7.6 may be used. An alternative is first to fit a polynomial response and then to use the methods of approximation theory to convert that into the desired form. In all cases, however, good choice of the centre of the design and the spacing of the levels is important for a succesful experiment.

When there are two or more factors with quantitative levels it may be very fruitful not merely to transform the component variables, but to define a linear transformation to new coordinates in the space of the factor variables. If, for instance, the response surface is approximately elliptical, new coordinates close to the principal axes of the ellipse will usually be helpful: a long thin ridge at an angle to the original coordinate axes would be poorly explored by a simple design without such a transformation of the x's. Of course to achieve a suitable transformation previous experimentation or theoretical analysis is needed. We shall suppose throughout the following discussion that any such transformation has already been used.

In many applications of factorial experiments the levels of the factors are defined by quantitative variables. In the discussion of Chapter 5 this information was not explicitly used, although the possibility was mentioned in Section 5.3.3.

We now suppose that all the factors of interest are quantitative, although it is straightforward to accommodate qualitative factors as well. In many cases, in the absence of a subject-matter basis for a specific nonlinear model, it would be reasonable to expect the response y to vary smoothly with the variables defining the factors; for example with two such factors we might assume

$$\begin{aligned} E(Y) &= \eta(x_1, x_2) = \beta_{00} + \beta_{10}x_1 + \beta_{01}x_2 \\ &+ \frac{1}{2}(\beta_{20}x_1^2 + 2\beta_{11}x_1x_2 + \beta_{02}x_2^2) \end{aligned} \quad (6.9)$$

with block and other effects added as appropriate. One interpretation of (6.9) is as two terms of a Taylor series expansion of $\eta(x_1, x_2)$ about some convenient origin.

In general, with k factors, the quadratic model for a response is

$$\begin{aligned} E(Y) = \eta(x_1, \ldots, x_k) &= \beta_{00\ldots} + \beta_{10\ldots}x_1 + \ldots + \beta_{0\ldots1}x_k \\ &+ \frac{1}{2}(\beta_{20\ldots}x_1^2 + 2\beta_{11\ldots}x_1x_2 + \ldots + \beta_{0\ldots2}x_k^2). \end{aligned} \quad (6.10)$$

A 2^k design has each treatment factor set at two levels, $x_i = \pm 1$, say. In Section 5.5 we used the values 0 and 1, but it is more convenient in the present discussion if the treatment levels are centred on zero. This design does not permit estimation of all the parameters in (6.10), as $x_i^2 \equiv 1$, so the coefficients of pure quadratic terms are confounded with the main effect. Indeed from observations at two levels it can hardly be possible to assess nonlinearity! However, the parameters $\beta_{10\ldots}$, $\beta_{01\ldots}$ and so on are readily identified with what in Section 5.5 were called main effects, i.e.

$$\begin{aligned} 2\hat{\beta}_{10\ldots} =\ & \text{average response at high level of factor 1} \\ & - \text{average response at low level of factor 1}, \end{aligned}$$

for example. Further, the cross-product parameters are identified with the interaction effects, $\beta_{11\ldots}$, for example, measuring the rate of change with x_2 of the linear regression of y on x_1.

In a fractional replicate of the full 2^k design, we can estimate linear terms $\beta_{10\ldots}$, $\beta_{01\ldots}$ and so on, as long as main effects are not

DESIGNS FOR QUANTITATIVE FACTORS

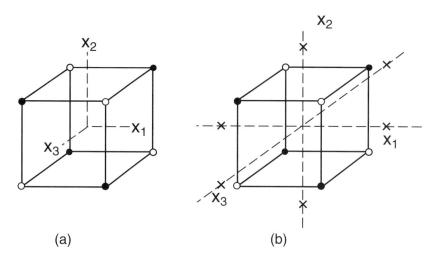

Figure 6.1 (a). Design space for three factor experiment. Full 2^3 indicated by vertices of cube. Closed and open circles, points of one-half replicates with alias $I = ABC$. (b) Axial points added to form central composite design.

aliased with each other. Similarly we can estimate cross-product parameters $\beta_{11...}$, etc., if two factor interactions are not aliased with any main effects.

To estimate the pure quadratic terms in the response, it is necessary to add design points at more levels of x_i. One possibility is to add the centre point $(0, \ldots, 0)$; this permits estimation of the sum of all the pure quadratic terms and may be useful when the goal is to determine the point of maximum or minimum response or to check whether a linear approximation is adequate against strongly convex or strongly concave alternatives.

Figure 6.1a displays the design space for the case of three factors; the points on the vertices of the cube are those used in a full 2^3 factorial. Two half fractions of the factorial are indicated by the use of closed or open circles. Either of these half fractions permits estimation of the main effects, β_{100}, β_{010} and β_{001}. Addition of one or more points at $(0, 0, 0)$ permits estimation of $\beta_{200} + \beta_{020} + \beta_{002}$; replicate centre points can provide an internal estimate of error, which should be compared to any error estimates available from external sources.

In order to estimate the pure quadratic terms separately, we must include points for at least three levels of x_i. One possibility is to use a complete or fractional 3^k factorial design. An alternative design quite widely used in industrial applications is the *central composite design*, in which a 2^k design or fraction thereof is augmented by one or more central points and by design points along the coordinate axes at $(\alpha, 0, \ldots, 0)$, $(-\alpha, 0, \ldots, 0)$ and so on. These axial points are added to the 2^3 design in Fig. 6.1b. One approach to choosing the coded value for α is to require that the estimated variance of the predicted response depends only on the distance from the centre point of the design space. Such designs are called *rotatable*. The criterion is, however, dependent on the scaling of the levels of the different factors; see Exercise 6.8.

6.6.2 Search for optima

Response surface designs are used, as their name implies, to investigate the shape of the dependence of the response on quantitative factors, and sometimes to determine the estimated position of maximum or minimum response, or more realistically a region in which close to optimal response is achieved. As at (6.10), this shape is often approximated by a quadratic, and once the coefficients are estimated the point of stationarity is readily identified. However if the response surface appears to be essentially linear in the range of x considered, and indeed whenever the formal stationary point lies well outside the region of investigation, further work will be needed to identify a stationary point at all satisfactorily. Extrapolation is not reliable as it is very sensitive to the quadratic or other model used.

In typical applications a sequence of experiments is used, first to identify important factors and then to find the region of maximum response. The method of steepest ascents can be used to suggest regions of the design space to be next explored, although scale dependence of the procedure is a major limitation. Typically the first experiment will not cover the region of optimality and a linear model will provide an adequate fit. The steepest ascent direction can be estimated from this linear model as the vector orthogonal to the fitted plane, although as noted above this depends on the relative units in which the x's are measured and this will usually be rather arbitrary.

DESIGNS FOR QUANTITATIVE FACTORS 153

6.6.3 Quality and quantity interaction

In most contexts the simple additive model provides a natural basis for the assessment of interaction. In special circumstances, however, there may be other possibilities, especially if one of the factors has quantitative levels. Suppose, for instance, that in a two factor experiment a level, i, of the first factor is labelled by a quantitative variable x_i, corresponding to the dose or quantity of some treatment, measured on the same scale for all levels j of the second factor which is regarded as qualitative.

One possible simple structure would arise if the difference in effect between two levels of j is proportional to the known level x_i, so that if Y_{ij} is the response in combination (i, j), then

$$E(Y_{ij}) = \alpha_j + \beta_j x_i, \qquad (6.11)$$

with the usual assumption about errors; that is, we have separate linear regressions on x_i for each level of the qualitative factor.

A special case, sometimes referred to as the interaction of quality and quantity, arises when at $x_i = 0$ we have that all factorial combinations are equivalent. Then $\alpha_j = \alpha$ and the model becomes

$$E(Y_{ij}) = \alpha + \beta_j x_i. \qquad (6.12)$$

Illustration. The application of a particular active agent, for example nitrogenous fertiliser, may be possible in various forms: the amount of fertiliser is the quantitative factor, and the variant of application the qualitative factor. If the amount is zero then the treatment is no additional fertiliser whatever the variant, so that all factorial combinations with $x_i = 0$ are identical.

In such situations it might be questioned whether the full factorial design, leading to multiple applications of the same treatment, is appropriate, although it is natural if a main effect of dose averaged over variants is required. With three levels of x_i, say 0, 1 and 2, and k levels of the second factor arranged in r blocks with $3k$ units per block the analysis of variance table will have the form outlined in Table 6.14.

Here there are two error lines, the usual one for a randomized block experiment and an additional one, shown last, from the variation within blocks between units receiving the identical zero treatment.

To interpret the treatment effect it would often be helpful to fit by least squares some or all of the following models:

$$E(Y_{ij}) = \alpha,$$
$$E(Y_{ij}) = \alpha + \beta x_i,$$
$$E(Y_{ij}) = \alpha + \beta_j x_i,$$
$$E(Y_{ij}) = \alpha + \beta x_i + \gamma x_i^2,$$
$$E(Y_{ij}) = \alpha + \beta_j x_i + \gamma x_i^2,$$
$$E(Y_{ij}) = \alpha + \beta_j x_i + \gamma_j x_i^2.$$

The last is a saturated model accounting for the full sum of squares for treatments. The others have fairly clear interpretations. Note that the conventional main effects model is not included in this list.

6.6.4 Mixture experiments

A special kind of experiment with quantitative levels arises when the factor levels $x_j, j = 1, \ldots, k$ represent the proportions of k components in a mixture. For all points in the design space

$$\Sigma x_j = 1, \qquad (6.13)$$

so that the design region is all or part of the unit simplex. A number of different situations can arise and we outline here only a few key ideas, concentrating for ease of exposition on small values of k.

First, one or more components may represent amounts of trace elements. For example, with $k = 3$, only very small values of x_1 may be of interest. Then (6.13) implies that $x_2 + x_3$ is effectively constant and in this particular case we could take x_1 and the proportion $x_2/(x_2 + x_3)$ as independent coordinates specifying treat-

Table 6.14 *Analysis of variance for a blocked design with treatment effects as in (6.11).*

Source	D.f.
Treatments	$2k$
Blocks	$r - 1$
Treatments× Blocks	$2k(r - 1)$
Error within Blocks	$r(k - 1)$

ments. More generally the dimension of the design space affected by the constraint (6.13) is $k-1$ minus the number of trace elements.

Next it will often happen that only treatments with all components present are of interest and indeed there may be quite strong restrictions on the combinations of components that are of concern. This means that the effective design space may be quite complicated; the algorithms of optimal design theory sketched in Section 7.4 may then be very valuable, especially in finding an initial design for more detailed study.

It is usually convenient to use simplex coordinates. In the case $k = 3$ these are triangular coordinates: the possible mixtures are represented by points in an equilateral triangle with the vertices corresponding to the pure mixtures $(1,0,0), (0,1,0)$ and $(0,0,1)$. For a general point (x_1, x_2, x_3), the coordinate x_1, say, is the area of the triangle formed by the point and the complementary vertices (0, 1, 0) and (0, 0, 1). The following discussion applies when the design space is the full triangle or, with minor modification, if it is a triangle contained within the full space.

At a relatively descriptive level there are two basic designs that in a sense are analogues of standard factorial designs. In the *simplex centroid* design, there are $2^k - 1$ distinct points, the k pure components such as $(1, 0, \ldots, 0)$, the $k(k-1)/2$ simple mixtures such as $(1/2, 1/2, 0, \ldots, 0)$ and so on up to the complete mixture $(1/k, \ldots, 1/k)$. Note that all components present are present in equal proportions. This may be contrasted with the *simplex lattice* designs of order (k, d) which are intended to support the fitting of a polynomial of degree d. Here the possible values of each x_j are $0, 1/d, 2/d, \ldots, 1$ and the design consists of all combinations of these values that satisfy the constraint $\Sigma x_j = 1$.

As already noted if the object is to study the behaviour of mixtures when one or more of the components are at very low proportions, or if singular behaviour is expected as one component becomes absent, these designs are not directly suitable, although they may be useful as the basis of a design for the other components of the mixture. Fitting of a polynomial response surface is unlikely to be adequate.

If polynomial fitting is likely to be sensible, there are two broad approaches to model parameterization, affecting analysis rather than design. In the first there is no attempt to give individual parameters specific interpretation, the polynomial being regarded as essentially a smoothing device for describing the whole surface.

The defining constraint $\Sigma x_j = 1$ can be used in various slightly different ways to define a unique parameterization of the model. One is to produce homogeneous forms. For example to produce a homogeneous expression of degree two we start with an ordinary second degree representation, multiply the constant by $(\Sigma x_j)^2$ and the linear terms by Σx_j leading to the general form

$$\Sigma_{i \leq j} \delta_{ij} x_i x_j, \tag{6.14}$$

with $k(k+1)/2$ independent parameters to be fitted by least squares in the usual way. Interpretation of single parameters on their own is not possible.

Other parameterizations are possible which do allow interpretation in terms of responses to pure mixtures, for example vertices of the simplex, simple binary mixtures, and so on.

A further possibility, which is essentially just a reparameterization of the first, is to aim for interpretable parameters in terms of contrasts and for this additional information must be inserted. One possibility is to consider a reference or standard mixture $(s_1, \ldots, s_k)$. The general idea is that to isolate the effect of, say, the first component we imagine x_1 increased to $x_1 + \Delta$. The other components must change and we suppose that they do so in accordance with the standard mixture, i.e. for $j \neq 1$, the change in x_j is to $x_j - \Delta s_j/(1 - s_1)$. Thus if we start from the usual linear model $\beta_0 + \Sigma \beta_j x_j$ imposition of the constraint

$$\Sigma \beta_j s_j = 0$$

will lead to a form in which a change Δ in x_1 changes the expected response by $\beta_1 \Delta/(1 - s_1)$. This leads finally to writing the linear response model in the form

$$\beta_0 + \Sigma \beta_j x_j/(1 - s_j) \tag{6.15}$$

with the constraint noted above. A similar argument applies to higher degree polynomials. The general issue is that of defining component-wise directional derivatives on a surface for which the simplex coordinate system is mathematically the most natural, but for reasons of physical interpretation not appropriate.

6.7 Taguchi methods

6.7.1 General remarks

Many of the ideas discussed in this book were first formulated in connection with agricultural field trials and were then applied in other areas of what may be broadly called biometry. Industrial applications soon followed and by the late 1930's factorial experiments, randomized blocks and Latin squares were quite widely used, in particular in the textile industries where control of product variability is of central importance. A further major development came in the 1950's in particular by the work of Box and associates on design with quantitative factors and with the search for optimum operating conditions in the process industries. Although first developed partly in a biometric context, fractional replication was first widely used in this industrial setting. The next major development came in the late 1970's with the introduction via Japan of what have been called Taguchi methods. Indeed in some discussions the term Taguchi design is misleadingly used as being virtually synonymous with industrial factorial experimentation.

There are several somewhat separate aspects to the so-called Taguchi method, which can broadly be divided into philosophical, design, and analysis. The philosophical aspects relate to the creation of working conditions conducive to the continuous emphasis on ensuring quality in production, and are related to the similarly motivated but more broad ranging ideas of Deming and to the notion of evolutionary operation.

We discuss here briefly the novel design aspects of Taguchi's contributions. One is the emphasis on the study and control of product variability, especially in contexts where achievement of a target mean value of some feature is relatively easy and where high quality hinges on low variability. Factors which cannot be controlled in a production environment but which can be controlled in a research setting are deliberately varied as so-called noise factors, often in split-unit designs. Another is the systematic use of orthogonal arrays to investigate main effects and sometimes two factor interactions.

The designs most closely associated with the Taguchi method are orthogonal arrays as described in Section 6.3, often Plackett-Burman two and three level arrays. There tends to be an emphasis in Taguchi's writing on designs for the estimation only of main effects; it is argued that in each experiment the factor levels can or should be chosen to eliminate or minimize the size of interactions among the controllable factors.

We shall not discuss some special methods of analysis introduced by Taguchi which are less widely accepted. Where product variability is of concern the analysis of log sample variances will often be effective.

The popularization of the use of fractional factorials and related designs and the emphasis on designing for reduction in variability and explicit accommodation of uncontrollable variability, although all having a long history, have given Taguchi's approach considerable appeal.

6.7.2 Example

This example is a case study from the electronics industry, as described by Logothetis (1990). The purpose of the experiment was to investigate the effect of six factors on the etch rate (in Å/min) of the aluminium-silicon layer placed on the surface of an integrated circuit. The six factors, labelled here A to F, control various conditions of manufacture, and three levels of each factor were chosen for the experiment. A seventh factor of interest, the over-etch time, was controllable under experimental conditions but not under manufacturing conditions. In this experiment it was set at two levels. Finally, the etch rate was measured at five fixed locations on each experimental unit, called a wafer: four corners and a centre point.

The design used for the six controllable factors is given in Table 6.15: it is an orthogonal array which in compilations of orthogonal array designs is denoted by $L_{18}(3^6)$ to indicate eighteen runs, and six factors with three levels each.

Table 6.16 shows the mean etch rate across the five locations on each wafer. The individual observations are given by Logothetis (1990). The two mean values for each factor combination correspond to the two levels of the "uncontrollable" factor, the over-etch rate. This factor has been combined with the orthogonal array in a split-unit design. The factor settings A up to F are assigned to whole units, and the two wafers assigned to different values of OE are the sub-units.

The design permits estimation of the linear and quadratic main effects of the six factors, and five further effects. All these effects are of course highly aliased with interactions. These five further

Table 6.15 *Design for the electronics example.*

A	B	C	D	E	F
−1	−1	−1	−1	−1	−1
−1	0	0	0	0	0
−1	1	1	1	1	1
0	−1	−1	0	0	1
0	0	0	1	1	−1
0	1	1	−1	−1	0
1	−1	0	−1	1	0
1	0	1	0	−1	1
1	1	−1	1	0	−1
−1	−1	1	1	0	0
−1	0	−1	−1	1	1
−1	1	0	0	−1	−1
0	−1	0	1	−1	1
0	0	1	−1	0	−1
0	1	−1	0	1	0
1	−1	1	0	1	−1
1	0	−1	1	−1	0
1	1	0	−1	0	1

effects are pooled to form an estimate of error for the main effects, and the analysis of variance table is as indicated in Table 6.17.

From this we see that the main effects of factors A, E and F are important, and partitioning of the main effects into linear and quadratic components shows that the linear effects of these factors predominate. This partitioning also indicates a suggestive quadratic effect of B. The AE linear by linear interaction is aliased with the linear effect of F and the quadratic effect of B, so the interpretation of the results is not completely straightforward. The simplest explanation is that the linear effects of A, E and AE are the most important influences on the etch rate.

The analysis of the subunits shows that the over-etch time does have a significant effect on the response, and there are suggestive interactions of this with A, B, D, and E. These interaction effects are much smaller than the main effects of the controllable factors. Note from Table 6.17 that the subunit variation between wafers is much smaller than the whole unit variation, as is often the case.

Table 6.16 *Mean etch rate (Å min^{-1}) for silicon wafers under various conditions.*

run	OE, 30s	OE, 90s	mean
1	4750	5050	4900
2	5444	5884	5664
3	5802	6152	5977
4	6088	6216	6152
5	9000	9390	9195
6	5236	5902	5569
7	12960	12660	12810
8	5306	5476	5391
9	9370	9812	9591
10	4942	5206	5074
11	5516	5614	5565
12	5108	5322	5210
13	4890	5108	4999
14	8334	8744	8539
15	10750	10750	10750
16	12508	11778	12143
17	5762	6286	6024
18	8692	8920	8806

6.8 Conclusion

In these six chapters we have followed a largely traditional path through the main issues of experimental design. In the following two chapters we introduce some more specialized topics. Throughout there is some danger that the key concepts become obscured in the details.

The main elements of good design may in our view be summarized as follows.

Experimental units are chosen; these are defined by the smallest subdivision of the material such that any two units may receive different treatments. A structure across different units is characterized, typically by some mixture of cross-classification and nesting and possibly baseline variables. The cross-classification is determined both by blocks (rows, columns, etc.) of no intrinsic interest and by strata determined by intrinsic features of the units (for

Table 6.17 *Analysis of variance for mean etch rate.*

	Source	Sum of sq. $(\times 10^6)$	D.f.	Mean sq. $(\times 10^6)$
Whole unit	A	84083	2	42041
	B	6997	2	3498
	C	3290	2	1645
	D	5436	2	2718
	E	98895	2	49448
	F	28374	2	14187
	Whole unit error	4405	2	881
Subunit	OE	408	1	408
	$OE \times A$	112	2	56
	$OE \times B$	245	2	122
	$OE \times C$	5.9	2	3.0
	$OE \times D$	159	2	79.5
	$OE \times E$	272	2	136
	$OE \times F$	13.3	2	6.6
	Subunit error	55.4	5	11.1

example, gender). Blocks are used for error control and strata to investigate possible interaction with treatments. Interaction of the treatment effects with blocks and variation among nested units is used to estimate error.

Treatments are chosen and possible structure in them identified, typically via a factorial structure of qualitative and quantitative factors.

Appropriate design consists in matching the treatment and unit structures to ensure that bias is eliminated, notably by randomization, that random error is controlled, usually by blocking, and that analysis appropriate to the design is achieved, in the simplest case via a linear model implicitly determined by the design, the randomization and a common assumption of unit-treatment additivity.

Broadly, in agricultural field trials structure of the units (plots) is a central focus, in industrial experiments structure of the treatments is of prime concern, whereas in most clinical trials a key issue is the avoidance of bias and the accrual of sufficient units (patients) to achieve adequate estimation of the relatively modest treatment differences commonly encountered. More generally each new field of application has its own special features; nevertheless common principles apply.

6.9 Bibliographic notes

The material in Sections 6.2, 6.3 and 6.4 stems largely from Yates (1935, 1937). It is described, for example, by Kempthorne (1952) and by Cochran and Cox (1958). Some of the more mathematical considerations are developed from Bose (1938).

Orthogonal arrays of strength 2, defined via Hadamard matrices, were introduced by Plackett and Burman (1945); the definition used in Section 6.3 is due to Rao (1947). Bose and Bush (1952) derived a number of upper bounds for the maximum possible number of columns for orthogonal arrays of strength 2 and 3, and introduced several methods of construction of orthogonal arrays that have since been generalized. Dey and Mukerjee (1999) survey the current known bounds and illustrate the various methods of construction, with an emphasis on orthogonal arrays relevant to fractional factorial designs. Hedayat, Sloane and Stufken (1999) provide an encyclopedic survey of the existence and construction of orthogonal arrays, their connections to Galois fields, error-correcting codes, difference schemes and Hadamard matrices, and their uses in statistics. The array illustrated in Table 6.6 is constructed in Wang and Wu (1991).

Supersaturated designs with the factor levels randomized, so-called random balance designs, were popular in industrial experimentation for a period in the 1950's but following critical discussion of the first paper on the subject (Satterthwaite, 1958) their use declined. Booth and Cox (1962) constructed systematic designs by computer enumeration. See Hurrion and Birgil (1999) for an empirical study.

Box and Wilson (1951) introduced designs for finding optimum operating conditions and the subsequent body of work by Box and his associates is described by Box, Hunter and Hunter (1978). Chapter 15 in particular provides a detailed example of sequential experimentation towards the region of the maximum, followed by the fitting of a central composite design in the region of the maximum. The general idea is that only main effects and perhaps a few

FURTHER RESULTS AND EXERCISES 163

two factor interactions are likely to be important. The detailed study of follow-up designs by Meyer, Steinberg and Box (1996) hinges rather on the notion that only a small number of factors, main effects and their interactions, are likely to play a major role.

The first systematic study of mixture designs and associated polynomial representations was done by Scheffé (1958), at the suggestion of Cuthbert Daniel, motivated by industrial applications. Earlier suggestions of designs by Quenouille (1953) and Claringbold (1955) were biologically motivated. A thorough account of the topic is in the book by Cornell (1981). The representation via a reference mixture is discussed in more detail by Cox (1971).

The statistical aspects of Taguchi's methods are best approached via the wide-ranging panel discussion edited by Nair (1992) and the book of Logothetis and Wynn (1989). For evolutionary operation, see Box and Draper (1969).

The example in Section 6.7 is discussed by Logothetis (1990), Fearn (1992) and Tsai et al. (1996). Fearn (1992) pointed out that the aliasing structure complicates interpretation of the results. The split plot analysis follows Tsai et al. (1996). The three papers are give many more details and a variety of approaches to the problem. There are also some informative interaction plots presented in the two latter papers. For an extended form of Taguchi-type designs for studying noise factors, see Rosenbaum (1999a).

Nelder (1965a, b) gives a systematic account of an approach to design and analysis that emphasizes treatment and unit structures as basic principles. For a recent elaboration, see Brien and Payne (1999).

6.10 Further results and exercises

1. A 2^4 experiment is to be run in 4 blocks with 4 units per block. Take as the generators ABC and BCD, thus confounding also the two factor interaction AD with blocks and display the treatments to be applied in each block. Now show that if it is possible to replicate the experiment 6 times, it is possible to confound each two factor interaction exactly once. Then show that 5/6 of the units give information about, say AB, and that if the ratio σ_c/σ is small enough, it is possible to estimate the two factor interactions more precisely after confounding, where σ_c^2 is the variance of responses within the same block and σ^2 is the variance of all responses.

2. Show that the 2^k experiment can be confounded in 2^{k-1} blocks of two units per block allowing the estimation of main effects from within block comparisons. Suggest a scheme of partial confounding appropriate if two factor interactions are also required.

3. Double confounding in 2^k: Let $u, v, \ldots$ and $x, y, \ldots$ be $r + c = k$ independent elements of the treatments group. Write out the $2^r \times 2^c$ array

$$
\begin{array}{ccccccc}
1 & x & y & xy & z & \ldots \\
u & ux & uy & uxy & uz & \\
v & vx & \ldots & & & \\
uv & & & & & \\
w & & & & & \\
\vdots & & & & &
\end{array}
$$

The first column is a subgroup and the other columns are cosets, i.e. there is a subgroup of contrasts confounded with columns, defined by generators $X, Y, \ldots$. Likewise there are generators $U, V, \ldots$ defining the contrasts confounded with rows. Show that $X, Y, \ldots; U, V, \ldots$ are a complete set of generators of the contrasts group.

4. We can formally regard a factor at four levels, $1, a, a^2, a^3$ as the product of two factors at two levels, by writing, for example 1, X, Y, and XY for the four levels. The three contrasts X, Y, and XY are three degrees of freedom representing the main effect of A. Often XY is of equal importance with X and Y and would be preserved in a system of confounding.

(a) Show how to arrange a 4×2^2 in blocks of eight with three replicates in a balanced design, partially confounding XBC, YBC and therefore also $XYBC$.

(b) If the four levels of the factor are equally spaced, express the linear, quadratic and cubic components of regression in terms of X, Y, and XY. Show that the Y equals the quadratic component and that if XY is confounded and the cubic regression is negligible, then X gives the linear component.

Yates (1937) showed how to confound the 3×2^2 in blocks of six, and the 4×2^n in blocks of $4 \times 2^{n-1}$ and $4 \times 2^{n-2}$. He also constructed the $3^n \times 2$ in blocks of $3^{n-1} \times 2$ and $3^{n-2} \times 2$. These designs are reproduced in many textbooks.

5. Discuss the connection between supersaturated designs and the solution of the following problem. Given $2m$ coins all but one of equal mass and one with larger mass and a balance with two pans thus capable of discriminating larger from smaller total masses, how many weighings are needed to find the anomalous coin.

By simulation or theoretical analysis examine the consequences in analysing data from the design of Table 6.7 of the presence of one, two, three or more main effects.

6. Explore the possibilities, including the form of the analysis of variance table, for designs of Latin square form in which in addition to the treatments constituting the Latin square further treatments are applied to whole rows and/or whole columns of the square. These will typically give contrasts for these further treatments of low precision; note that the experiment is essentially of split plot form with two sets of whole unit treatments, one for rows and one for columns. The designs are variously called plaid designs or criss-cross designs. See Yates (1937) and for a discussion of somewhat related designs applied to an experiment on medical training for pain assessment, Farewell and Herzberg (2000).

7. Suppose that in a split unit experiment it is required to compare two treatments with different levels of both whole unit and subunit treatments. Show how to estimate the standard error of the difference via a combination of the two residual mean squares. How would approximate confidence limits for the difference be found either by use of the Student t distribution with an approximate number of degrees of freedom or by a likelihood-based method?

8. In a response surface design with levels determined by variables $x_1, \ldots, x_k$ the variance of the estimated response at position x under a given model, for example a polynomial of degree d, can be regarded as a function of x. If the contours of constant variance are spherical centred on the origin the design is called rotatable; see Section 6.6.1. Note that the definition depends not merely on the choice of origin for x but more critically on the relative units in which the different x's are measured. For a

quadratic model the condition for rotatability, taking the centroid of the design points as the origin, requires all variables to have the same second and fourth moments and $\Sigma x_{iu}^4 = 3\Sigma x_{iu}^2 x_{ju}^2$ for all $i \neq j$.

Show that for a quadratic model with 2^k factorial design points $(\pm 1, \ldots, \pm 1)$ and $2k$ axial points $(\pm a, 0, \ldots), \ldots, (0, \ldots, \pm a)$, the design is rotatable if and only if $a = (\sqrt{2})^k$. For comparative purposes it is more interesting to examine differences between estimated responses at two points x', x'', say. It can be shown that in important special cases rotatability implies that the variance depends only on the distances of the points from the origin and the angle between the corresponding vectors. Rotatability was introduced by Box and Hunter (1957) and the discussion of differences is due to Herzberg (1967).

9. The treatment structure for the example discussed in Section 4.2.6 was factorial, with three controllable factors expected to affect the properties of the response. These three factors were quantitative, and set at three equally spaced levels, here shown in coded values, following a central composite design. Each of the eight factorial points $(\pm 1, \pm 1, \pm 1)$ were used twice, the centre point $(0, 0, 0)$ was replicated six times, and the six axial points $(\pm 1, 0, 0)$, $(0, \pm 1, 0)$, and $(0, 0, \pm 1)$ were used once. The data and treatment assignment to blocks are shown in Table 4.13; Table 6.18 shows the factorial points corresponding to each of the treatments.

A quadratic model in x_A, x_B, and x_C has nine parameters in addition to the overall mean. Fit this model, adjusted for blocks, and discuss how the linear and quadratic effects of the three factors may be estimated. What additional effects may be estimated from the five remaining treatment degrees of freedom? Discuss how the replication of the centre point in three different blocks may be used as an adjunct to the estimate of error obtained from Table 4.15.

Gilmour and Ringrose (1999) discuss the data in the light of fitting response surface models. Blocking of central composite designs is discussed in Box and Hunter (1957); see also Dean and Voss (1999, Chapter 16).

10. What would be the interpretation in the quality-quantity example of Section 6.6.3 if the upper of the two error mean squares

Table 6.18 *Factorial treatment structure for the incomplete block design of Gilmour and Ringrose (1999); data and preliminary analysis are given in Table 4.13.*

Trtm	1	2	3	4	5	6	7	8
x_A	−1	−1	−1	−1	1	1	1	1
x_B	−1	−1	1	1	−1	−1	1	1
x_C	−1	1	−1	1	−1	1	−1	1
Day	1, 6	3, 7	3, 5	2, 6	2, 3	4, 6	6, 7	1, 3

Trtm	9	10	11	12	13	14	15
x_A	0	−1	0	0	1	0	0
x_B	0	0	−1	0	0	1	0
x_C	0	0	0	−1	0	0	1
Day	1, 2, 7	4	5	4	5	4	5

were to be much larger (or smaller) than the lower? Compare the discussion in the text with that of Fisher (1935, Chapter 8).

11. Show that for a second degree polynomial for a mixture experiment a canonical form different from the one in the text results if we eliminate the constant term by multiplying by Σx_j and eliminate the squared terms such as x_j^2 by writing them in the form $x_j(1 - \Sigma_{k \neq j} x_k)$. Examine the extension of this and other forms to higher degree polynomials.

12. In one form of analysis of Taguchi-type designs a variance is calculated for each combination of fixed factors as between the observations at different levels of the noise factors. Under what exceptional special conditions would these variances have a direct interpretation as variances to be empirically realized in applications? Note that the distribution of these variances under normal-theory assumptions has a noncentral chi-squared form. A standard method of analyzing sets of normal-theory estimates of variance with d degrees of freedom uses the theoretical variance of approximately $2/d$ for the log variances and a multiplicative systematic structure for the variances. Show that this would tend to underestimate the precision of the conclusions.

13. Pistone and Wynn (1996) suggested a systematic approach to the fitting of polynomial and some other models to essentially

arbitrary designs. A key aspect is that a design is specified via polynomials that vanish at the design points. For example, the 2^2 design with observations at $(\pm 1, \pm 1)$ is specified by the simultaneous equations $x_1^2 - 1 = 0, x_2^2 - 1 = 0$. A general polynomial in (x_1, x_2) can then be written as

$$k_1(x_1, x_2)(x_1^2 - 1) + k_2(x_1, x_2)(x_2^2 - 1) + r(x_1, x_2),$$

where $r(x_1, x_2)$ is a linear combination of $1, x_1, x_2, x_1 x_2$ and these terms specify a saturated model for this design. More generally a design with n distinct points together with an ordering of the monomial expressions $x_1^{a_1} \cdots x_k^{a_k}$, in the above example, $1 \prec x_1 \prec x_2 \prec x_1 x_2$, determines a *Gröbner basis*, which is a set of polynomials $\{g_1, \ldots, g_m\}$ such that the design points satisfy the simultaneous equations $g_1 = 0, \ldots, g_m = 0$. Moreover when an arbitrary polynomial is written

$$\sum k_s(x) g_s(x) + r(x),$$

the remainder $r(x)$ specifies a saturated model for the design respecting the monomial ordering. Computer algorithms for finding Gröbner bases are available. Once constructed the terms in the saturated model are found via monomials not divisible by the leading terms of the bases. For a full account, see Pistone, Riccomagno and Wynn (2000).

CHAPTER 7

Optimal design

7.1 General remarks

Most of the previous discussion of the choice of designs has been on a relatively informal basis, emphasizing the desirability of generally plausible requirements of balance and the closely associated notion of orthogonality; see Section 1.7. We now consider the extent to which the design process can be formalized and optimality criteria used to deduce a design. We give only an outline of what is a quite extensive theoretical development.

This theory serves two rather different purposes. One is to clarify the properties of established designs whose good properties have no doubt always been understood at a less formal level. The other is to give a basis for suggesting designs in nonstandard situations.

We begin with a very simple situation that will then serve to illustrate the general discussion and in particular a key result, the General Equivalence Theorem of Kiefer and Wolfowitz.

7.2 Some simple examples

7.2.1 Straight line through the origin

Suppose that it is possible to make n separate observations on a response variable Y whose distribution depends on a single explanatory variable x and that for each observation the investigator may choose a value of x in the closed interval $[-1, 1]$. We call the interval the design region $\mathcal{D}$ for the problem. Suppose further that values of Y for different individuals are uncorrelated of constant variance σ^2 and have

$$E(Y) = \beta x, \tag{7.1}$$

where β is an unknown parameter.

We thus have a very explicit formulation both of a model and a design restriction; the latter is supposed to come either from a practical constraint on the region of "safe" or accessible experimentation or from the consideration that the region is the largest over which the model can plausibly be used.

We specify the design used, i.e. the set of values $\{x_1,\ldots,x_n\}$ employed, by a measure $\xi(\cdot)$ over the design region attaching design mass $1/n$ to the relevant x for each observation; note that the same point in $\mathcal{D}$ may be used several times and then receives mass $1/n$ from each occurrence.

It is convenient to write

$$n^{-1}\Sigma x_i^2 = \int x^2 \xi(dx) = M(\xi), \tag{7.2}$$

say. We call $M(\xi)$ the *design second moment*; in general it is proportional to the Fisher information. If we analyse the responses by least squares

$$\text{var}(\hat{\beta}) = (\sigma^2/n)\{M(\xi)\}^{-1}. \tag{7.3}$$

To estimate the expected value of Y at an arbitrary value of $x \in \mathcal{D}$, say x_0, we take

$$\hat{Y}_0 = \hat{\beta}x_0, \tag{7.4}$$

with

$$\text{var}(\hat{Y}_0) = \frac{\sigma^2}{n}\{M(\xi)\}^{-1}x_0^2 = \frac{\sigma^2}{n}d(x_0,\xi), \tag{7.5}$$

say.

There are two types of optimality requirement that might now be imposed. One is to minimize $\text{var}(\hat{\beta})$ and this requires the maximization of $M(\xi)$. Any design which attaches all the design mass to the points ± 1 achieves this. Alternatively we may minimize the maximum over x_0 of $\text{var}(\hat{Y}_0)$. We have that

$$\bar{d}(\xi) = \sup_{x_0 \subset \mathcal{D}} d(x_0,\xi) = \{M(\xi)\}^{-1} \tag{7.6}$$

and this is minimized as before by maximizing $M(\xi)$. Further when this is done

$$\inf_\xi \bar{d}(\xi) = 1. \tag{7.7}$$

7.2.2 Straight line with intercept

Now consider the more general straight line model with an intercept,

$$E(Y) = \beta_0 + \beta_1 x. \tag{7.8}$$

SOME SIMPLE EXAMPLES

We generalize the design moment to the design moment matrix

$$M(\xi) = \begin{pmatrix} 1 & \bar{x} \\ \bar{x} & n^{-1}\Sigma x_i^2 \end{pmatrix}, \quad (7.9)$$

where $\bar{x} = n^{-1}\Sigma x_i$, for example, can be written as

$$\bar{x} = \int x\xi(dx). \quad (7.10)$$

The determinant and inverse of $M(\xi)$ are

$$\det M(\xi) = n^{-1}\Sigma(x_i - \bar{x})^2, \quad (7.11)$$

$$\{M(\xi)\}^{-1} = n\{\Sigma(x_i - \bar{x})^2\}^{-1} \begin{pmatrix} n^{-1}\Sigma x_i^2 & -\bar{x} \\ -\bar{x} & 1 \end{pmatrix}. \quad (7.12)$$

Now the covariance matrix of the estimated regression coefficients is $\{M(\xi)\}^{-1}\sigma^2/n$. It follows that, at least for even values of n, the determinant of M and the value of $\text{var}(\hat{\beta}_1)$ are minimized by putting design mass of $1/2$ at the points ± 1, i.e. spacing the points as far apart as allowable. There is a minor complication if n is odd.

Also the variance of $\hat{Y}_0 = \hat{\beta}_0 + \hat{\beta}_1 x_0$, the estimated mean response at the point x_0, is again $d(x_0, \xi)\sigma^2/n$, where now

$$d(x, \xi) = (1 \ \ x)\{M(\xi)\}^{-1}(1 \ \ x)^T.$$

It is convenient sometimes to regard this as defining the variance of prediction although for predicting a single outcome σ^2 would have to be added to the variance of the estimated mean response.

A direct calculation shows that for the symmetrical two-point design

$$\inf_\xi \bar{d}(\xi) = \inf_\xi \sup_{x \subset \mathcal{D}} d(x, \xi) = 2. \quad (7.13)$$

Again in this case, although not in general, different optimality requirements can be met simultaneously. In summary, any design with $\bar{x} = 0$ minimizes $\text{var}(\hat{\beta}_0)$, the design with equal mass $1/2$ at ± 1 minimizes $\text{var}(\hat{\beta}_1)$ and also can be shown to minimize $\bar{d}(\xi)$.

7.2.3 Critique

These results depend heavily on the precise specification of the model and of the design region. In many situations it would be a fatal criticism of the above design that it offers no possibility of checking the assumed linearity of the regression function; such checking does not feature in the optimality criteria used above so that it is no surprise that it does not feature in the optimal design.

We now investigate informally two approaches to the inclusion of some check of linearity. One is to use the optimal design for a proportion $(1-w)$ of the observations and to take the remaining proportion w at some third point, most naturally at zero in the absence of special considerations otherwise; see Section 6.6. The variance of the estimated slope is increased by a factor $(1-w)^{-1}$ and nonlinearity can be studied via the statistic comparing the sample mean Y at $x = 0$ with the mean of the remaining observations. That analysis is closely associated with the fitting of the quadratic model

$$E(Y) = \beta_0 + \beta_1 x + \beta_2 x^2. \tag{7.14}$$

A direct calculation shows that the optimal design for estimating β_1 has $w = 0$, the optimal design for estimating β_2 has $w = 1/2$ and that for minimizing both the determinant of the covariance matrix of β and the prediction based criterion $\bar{d}(\xi)$ has $w = 1/3$, leading also to

$$\inf_\xi \bar{d}(\xi) = \inf_\xi \sup_{x \subset \mathcal{D}} d(x, \xi) = 3. \tag{7.15}$$

If a suitable criterion balancing the relative importance of estimating β_1 and testing the adequacy of linearity were to be formulated then and only then could a suitable w be deduced within the present formulation.

7.2.4 Space-filling designs

There is another more extreme approach to design in this context. If a primary aspect were the exploration of an unknown function rather than the estimation of a slope then a reasonable strategy would be to spread the design points over the design region, for example approximately uniformly. It is a convenient approximation to allow the measure $\xi(\cdot)$ to be continuous and in particular to consider the uniform distribution on $(-1, 1)$. For the simple two-parameter linear model this would lead to the moment matrix

$$M(\xi) = \text{diag}(1, 1/3), \{M(\xi)\}^{-1} = \text{diag}(1, 3),$$

representing a three-fold increase in variance of the estimated slope as compared with the optimal design. Hardly surprisingly, the ob-

SOME GENERAL THEORY

jectives of estimating a parameter in a tightly specified model and of exploring an essentially unknown function lead to quite different designs. See Section 7.7 for an introduction to space-filling designs.

7.3 Some general theory

7.3.1 Formulation

We now sketch a more general formulation. The previous section provides motivation and exemplification of most of the ideas involved.

We consider a design region $\mathcal{D}$, typically a closed region in some Euclidean space, R^d, and a linear model specifying for a particular $x \subset \mathcal{D}$ that

$$E(Y) = \beta^T f(x), \tag{7.16}$$

where β is a $p \times 1$ vector of unknown parameters and $f(x)$ is a $p \times 1$ vector of known functions of x, for example the powers $\{1, x, x^2, \ldots, x^{p-1}\}$ or orthogonalized versions thereof, or the first few sinusoidal functions as the start of an empirical Fourier series.

We make the usual second moment error assumptions leading to the use of least squares estimates. Under some circumstances this might be justified by randomization.

A design is specified by an initially arbitrary measure $\xi(\cdot)$ assigning unit mass to the design region $\mathcal{D}$. If this is formed from atoms of size equal to $1/n$, where n is the specified number of observations, the design can be exactly realized in n observations, but otherwise the specification has to be regarded as an approximation valid in some sense for large n.

We define the moment matrix by

$$M(\xi) = \int f(x)f(x)^T \xi(dx), \tag{7.17}$$

so that the covariance matrix of the least squares estimate of β from n observations with variance σ^2 is

$$(\sigma^2/n)\{M(\xi)\}^{-1} \tag{7.18}$$

and the variance of the estimated mean response at point x is $(\sigma^2/n)d(x, \xi)$, where

$$d(x, \xi) = f(x)^T \{M(\xi)\}^{-1} f(x). \tag{7.19}$$

We define a design ξ^* to be *D–optimal* if it maximizes $\det M(\xi)$ or, of course equivalently minimizes $\det\{M(\xi)\}^{-1}$, the latter determining the generalized variance of the least squares estimate of β.

We define a design to be *G–optimal* if it minimizes

$$\bar{d}(\xi) = \sup_{x \subset \mathcal{D}} d(x,\xi). \qquad (7.20)$$

7.3.2 General equivalence theorem

A central result in the theory of optimal design, the General Equivalence Theorem, asserts that the design ξ^* that is *D*–optimal is also *G*–optimal and that

$$\bar{d}(\xi^*) = p, \qquad (7.21)$$

the number of parameters. The specific example of linear regression in Section 7.2 illustrates both parts of this.

We shall not give a full proof which requires showing overall optimality and uses some arguments in convex analysis. We will outline a proof of local optimality. For this we perturb the measure ξ to $(1-\epsilon)\xi + \epsilon \delta_x$, where δ_x is a unit atom at x. For a scalar, vector or matrix valued function $H(\xi)$ determined by the measure ξ, we can define a derivative

$$\mathrm{der}\{H(\xi), x\} = \lim_{\epsilon \to 0^+} \epsilon^{-1} H\{(1-\epsilon)\xi + \epsilon \delta_x\}, \qquad (7.22)$$

where the limit is taken as ϵ tends to zero through positive values. This is a generalization of the notion of partial or directional derivatives and a special case of Gâteaux derivatives, the last having a similar definition with more general perturbing measures. A necessary condition for ξ^* to produce a local stationary point in H is that $\mathrm{der}\{H(\xi^*), x\} = 0$ for all $x \in \mathcal{D}$.

To apply this to *D*–optimality we take H to be $\log \det M(\xi)$. For any nonsingular matrix A and any matrix B, we have that as ϵ tends to zero

$$\begin{aligned}\det(A + \epsilon B) &= \det(A)\det(I + \epsilon A^{-1} B) \\ &= \det(A)\{1 + \epsilon \mathrm{tr}(A^{-1} B)\} + O(\epsilon^2),\end{aligned} \qquad (7.23)$$

where tr denotes the trace of a square matrix. That is, at $\epsilon = 0$, we have that

$$(d/d\epsilon) \log \det(A + \epsilon B) = \mathrm{tr}(A^{-1} B). \qquad (7.24)$$

SOME GENERAL THEORY

The moment matrix $M(\xi)$ is linear in the measure ξ and if all the mass is concentrated at x the moment matrix would be $f(x)f^T(x)$. Thus

$$M\{(1-\epsilon)\xi + \epsilon\delta_x\} = M(\xi) + \epsilon\{f(x)f(x)^T - M(\xi)\}. \qquad (7.25)$$

Therefore

$$\text{der}[\log\det\{M(\xi),x\}] = \text{tr}\{M^{-1}f(x)f(x)^T - I\}. \qquad (7.26)$$

Note that the derivative is meaningful only as a right-hand derivative unless there is an atom at x in the measure ξ in which case the subtraction of a small atom is possible and negative ϵ allowable.

Now the trace of the first matrix on the right-hand side is equal to $\text{tr}\{f(x)^T M^{-1}f(x)\}$ and that of the second matrix is p, the dimension of the parameter vector:

$$\text{der}[\log\det\{M(\xi),x\}] = d(x,\xi) - p. \qquad (7.27)$$

Next suppose that we have a design measure ξ^* such that at all sets of points in $\mathcal{D}$ which have design measure zero, $d(x,\xi^*) < p$ and at all points with positive measure, including especially points with atomic design mass, $d(x,\xi^*) = p$. Then the design is locally D–optimal. For a perturbation that adds a small design mass where there was none before decreases the determinant whereas with respect to changes at other points there is a stationary point, in fact a local maximum. Global D–optimality and the second property of G–optimality hinge on convexity, G–optimality in particular on the generalized derivative being nonpositive, so that indeed the "worst" points for prediction are the points of positive mass where $\bar{d}(\xi^*) = p$.

The most important point to emerge from the above outline is the mathematical basis for the connection between the covariance matrix of the least squares estimates and the variance of estimated mean response and the origin of the, at first sight mysterious, identity between the maximum variance of prediction and the number of parameters.

7.3.3 Some special cases

If the D–optimal design has support on p distinct points with design masses $\delta_1,\ldots,\delta_p$ it follows that the moment matrix has the form

$$M(\xi) = C\text{diag}(\delta_1,\ldots,\delta_p)C^T, \qquad (7.28)$$

where the $p \times p$ matrix C depends only on the positions of the points. It follows that

$$\det M(\xi) = \{\det(C)\}^2 \Pi \delta_i. \tag{7.29}$$

The condition for D–optimality thus requires that $\Pi \delta_i$ is maximized subject to $\Sigma \delta_i = 1$ and this is easily shown to imply that all points have equal design mass $1/p$. Not all D–optimal designs are supported on as few as p points, however.

As an example we take the quadratic model of (7.14), and for convenience reparameterize in orthogonalized form as

$$E(Y_x) = \gamma_0 + \gamma_1 x + \gamma_2 (3x^2 - 2). \tag{7.30}$$

With equal design mass $1/3$ at $\{-1, 0, 1\}$ we have that

$$M(\xi^*) = \operatorname{diag}(1, 2/3, 2),$$
$$M^{-1}(\xi^*) = \operatorname{diag}(1, 3/2, 1/2), \tag{7.31}$$

so that on calculating $\operatorname{var}(\hat{Y}_0)$, we have that

$$d(x_0, \xi^*) = 3 - 9(x_0^2 - x_0^4)/2 \tag{7.32}$$

and all the properties listed above can be directly verified. That is, in the design region $\mathcal{D}$ the generalized derivative is negative at all points except the three points with positive support where it is zero and where the maximum standardized variance of prediction of 3 is achieved.

7.4 Other optimality criteria

The discussion in the previous section hinges on a link between a global criterion about the precision of the vector estimate $\hat{\beta}$ and the variance of prediction. A number of criteria other than D–optimality may be more appropriate. One important possibility is based on partitioning β into two components β_1 and β_2 and focusing on β_1 as the component of interest; in a special case β_1 consists of a single parameter.

If the information matrix of the full parameter vector is partitioned conformally as

$$M(\xi) = \begin{pmatrix} M_{11} & M_{12} \\ M_{21} & M_{22} \end{pmatrix} \tag{7.33}$$

the covariance matrix of the estimate of interest is proportional to the inverse of the matrix

$$M_{11.2}(\xi) = M_{11} - M_{12}M_{22}^{-1}M_{21} \qquad (7.34)$$

and we call a design D_s-*optimal* if the determinant of (7.34) is maximized.

An essentially equivalent but superficially more general formulation concerns the estimation of the parameter $A\beta$, where A is a $q \times p$ matrix with $q < p$; the corresponding notion may be called D_A-optimality.

Other notions that can be appropriate in special contexts include *A-optimality* in which the criterion to be minimized is $\text{tr}(M^{-1})$, and *E-optimality* which aims to minimize the variance of the least well-determined of a set of contrasts.

7.5 Algorithms for design construction

The use of the theory developed so far in this chapter is partly to verify that designs suggested from more qualitative considerations have good properties and partly to aid in the construction of designs for nonstandard situations. For example an unusual model may be fitted or a nonstandard design region may be involved.

There are two slightly different settings requiring use of optimal design algorithms. In one some observations may already be available and it is required to supplement these so that the full design is as effective as possible. In the second situation only a design region and a model are available. If an informative prior distribution were available and if this were to be used in the analysis as well as the choice of a design the two situations are essentially equivalent.

We shall concentrate on D–optimality. There are various algorithms for finding an optimal design; all are variants on the following idea. Start with an initial design, ξ_0, in the first problem above with that used in the first part of the experiment and in the second problem often with some initially plausible atomic arrangement with atoms of amount $1/N$, where N is large compared with the number n of points to be used. Cover the design region with a suitable network, $\mathcal{N}$, of points and compute the function $d(x, \xi_0)$ for all x in $\mathcal{N}$. The network $\mathcal{N}$ should be rich enough to contain close approximations to the points likely to have positive mass in the ultimate design. The idea is to add design mass where d is large until the D–optimality criterion is reasonably closely satisfied. Then it will be necessary to look at the design realizable with n observations and, especially importantly, to check that the design has no features undesirable in the light of some aspect not covered in the formal optimality criterion used.

For the construction of a design without previous observations one appealing formulation is as follows: at the kth step let the design measure be ξ_k. Remove an atom from the point in $\mathcal{N}$ with smallest $d(x, \xi_k)$ and attach it to the point with largest $d(x, \xi_k)$. There are many ways of accelerating such an algorithm.

In some ways the simplest algorithm, and one for which convergence can be proved, is at the kth step to find the point x_k at which $d(x, \xi_k)$ is maximized and to define

$$\xi_{k+1} = k\xi_k/(k+1) + \delta_{x_k}/(k+1). \tag{7.35}$$

If other optimality requirements are used a similar algorithm can be used based on the relevant directional derivative.

These algorithms give the optimizing ξ. The construction of the optimum n point design for given n is a combinatorial optimization problem and in general is much more difficult.

7.6 Nonlinear design

The above discussion concerns designs for problems in which analysis by least squares applied to linear models is appropriate. While quite strong specification is needed to deduce an optimal design the solution does not depend on the unknown parameter under study. When nonlinear models are appropriate for analysis, much of the previous theory applies replacing least squares estimation by maximum likelihood estimation but the optimal design will typically depend on the unknown parameter under study.

This means that either one must be content with optimality at the best prior estimate of the unknown parameter, checking for sensitivity to errors in this prior estimate, or, preferably, that a sequential approach is used.

As an illustration we discuss what was historically one of the first problems of optimal design to be analysed, the dilution series. Suppose that organisms are distributed at random at a rate μ per unit volume. If a unit volume is sampled the number of organisms will have a Poisson distribution of mean μ. If the unit volume is diluted by a factor k the number will have a Poisson distribution of mean μ/k. In some contexts it is much easier to check for presence or absence of the organism than it is to count numbers, leading at

NONLINEAR DESIGN

dilution k to consideration of a binary variable Y_k, where

$$P(Y_k = 0; k, \mu) = e^{-\mu/k}, \quad P(Y_k = 1; k, \mu) = 1 - e^{-\mu/k}, \quad (7.36)$$

corresponding respectively to the occurrence of no organism or one or more organisms.

A commonly used procedure is to examine r samples at each of a series of dilutions, for example taking $k = 1, 2, 4, \ldots$. The number of positive samples at a given k will have a binomial distribution and hence a log likelihood function can be found and μ estimated by maximum likelihood. Samples at both large and very small values of μ/k provide little quantitative information about μ; it is the samples with a value of k approximately equal to μ that are most informative. There is one unknown parameter and the design region is the set of allowable k, essentially all values greater than or equal to one. The design criterion is to minimize the asymptotic variance of the maximum likelihood estimate $\hat\mu$ or equivalently to maximize the Fisher information about μ. It is plausible and can be formally proved that this is done by choosing a single value of k.

The log likelihood for one observation on Y_k is

$$-\mu k^{-1}(1 - Y_k) + Y_k \log(1 - e^{-\mu/k}), \quad (7.37)$$

so that the expected information about μ is

$$\nu^2 e^{-\nu}(1 - e^{-\nu})^{-1}\mu^{-2}, \quad (7.38)$$

where $\nu = \mu/k$, the function having its maximum for given μ at $\nu = 1.594$, i.e. at $k_{\text{opt}} = 0.627\mu$, the corresponding expected information per observation about μ being $0.648/\mu^2$. If we made one direct count of the number of organisms at dilution k^* yielding a Poisson distributed observation of mean μ/k^* the expected information about μ is $1/(k^*\mu)$ and if it so happened that the above optimal dilution had been chosen, so that $k^* = k_{\text{opt}}$, the information about μ would then be $1.594/\mu^2$. Some of the loss of information involved in the simple dilution series method could be recovered if it were possible to record the response variable in the extended form 0, 1, more than 1. Then, of course, the optimality calculation would need revision. If particular interest was focused on whether μ is larger or smaller than some special value μ_0 it would be reasonable to use the design locally optimal at μ_0.

The key point is that the optimal choice of design depends on the unknown parameter μ and this is typical of nonlinear problems in general. To examine sensitivity to the choice of the dilution constant we write $k = ck_{\text{opt}}$, so that c is the ratio of the dilution used to its optimal value. The ratio of the resulting information to its optimal value can be found from (7.38). The dependence on errors in assessing k is asymmetric, the ratio being 0.804 at $c = 2$ and 0.675 at $c = 1/2$ and being 0.622 at $c = 3$ and 0.298 at $c = 1/3$. It would thus be better to dilute by too much than by too little if a single dilution strategy were to be adopted.

In normal theory regression problems there would be a similar dependence if the error variance depended on the unknown parameter controlling the expected value.

If the dependence of intrinsic precision on the unknown parameter is slight then designs virtually independent of the unknown parameter are achieved. For example, suppose that a reasonable model for the observed responses is of the generalized linear regression form in which $Y_1, \ldots, Y_n$ are independently distributed in the exponential family distribution with the density of Y_j being

$$\exp\{\phi_j y_j + a(y_j) - k(\phi_j)\},$$

where ϕ_j is the canonical parameter, $a(y_j)$ a normalizing function and $k(\phi_j)$ the cumulant function. Suppose also that for some known function $h(\cdot)$ the vector $h(\phi)$ with components $h(\phi_j)$ obeys the linear model

$$h(\phi) = \beta^T f(x), \qquad (7.39)$$

where β and $f(x)$ have the same interpretation as in the linear model (7.16). The natural analogue to the moment matrix is the information matrix for β,

$$\Sigma m\{\beta^T f(x_j)\} f(x_j) f(x_j)^T, \qquad (7.40)$$

where the function $m(\cdot)$ depends on the functions $h(\cdot)$ and $k(\cdot)$. If the dependence of the response on the covariates is weak, that is if we can work locally near $\beta = 0$, Taylor expansion shows that the information matrix is proportional to the moment matrix $M(\xi)$, so that, for example, the D–optimal design is the same as in the linear least squares case. This conclusion is qualitatively obvious when $h(\phi) = \phi$ on remarking that the canonical statistics are the same as in the normal-theory case and the dependence of their asymptotic distribution on β is by assumption weak. For example, locally at least near a full null hypothesis, optimal designs for linear logistic regression are the same as in the least squares case.

7.7 Space-filling designs

The optimal designs described in Sections 7.2–7.4 typically sample at the extreme points of the design space, as they are in effect targetted at minimizing the variance of prediction for a given model. In problems where the model is not well specified a preferable strategy is to take observations throughout the range of the design space, at least until more information about the shape of the response surface is obtained. Such designs are called *space-filling designs*.

In Section 6.6 we discussed two and three level factorial systems for exploration of response surfaces that were well approximated by quadratic polynomials. If the dependence of Y on $x_1, \ldots, x_k$ is highly nonlinear, either because the system is very complex, or the range of values of $\mathcal{X}$ is relatively large, then this approximation may be quite poor. A space-filling design is then useful for exploring the nature of the response surface.

Illustrations. In an engineering context, an experiment may involve running a large simulation of a complex deterministic system with given values for a number of tuning parameters. For example, in a fluid dynamics experiment, the system may be determined via the solution of a number of partial differential equations. As the system is deterministic, random error is unimportant. Of interest is the choice of tuning parameters needed to ensure that the simulator reproduces observations consistent with data from the physical system. Since each run of the simulator may be very expensive, efficient choice of the test values of the tuning parameters is important.

In modelling a stochastic epidemic there may be a number of parameters introduced to describe rates of infectivity, transmission under various mechanisms, and so on. At least in the early stages of the epidemic there will not be enough data to permit estimation of these parameters. Some progress can be made by simulating the model over the full range of parameter values, and accepting as plausible those parameter combinations which give results consistent with the available data. A space-filling design may be used to choose the parameter values for simulation of the model.

A commonly used space-filling design, called a *Latin hypercube*, is constructed as follows. The range for each of m design variables (tuning parameters in the illustrations above) is divided into n subintervals, equally spaced on the appropriate scale for each variable. An array of n rows and m columns is constructed by assigning to each column a random permutation of $\{1, \ldots, n\}$, independently of the other columns. Each row of the resulting array defines a design point for the n-run experiment, either at a fixed point in each subinterval, such as the endpoint or midpoint, or randomly sampled within the subinterval. These designs are generalizations of the lattice square, which is constructed from sets of orthogonal Latin squares in Section 8.5.

For example, with $n = 10$ and $m = 3$, we might obtain the array

$$\begin{pmatrix} 10 & 8 & 9 & 7 & 6 & 5 & 3 & 1 & 4 & 2 \\ 9 & 2 & 3 & 8 & 5 & 1 & 4 & 10 & 6 & 7 \\ 5 & 2 & 1 & 7 & 6 & 10 & 9 & 8 & 4 & 3 \end{pmatrix}^T. \quad (7.41)$$

Thus if each of the three design variables take values equally spaced on $(0, 1)$, and we sample midpoints, the design points on $(0, 1)^3$ are $(0.45, 0.85, 0.95), (0.15, 0.15, 0.75)$, and so on. With $n = 9$ we could use the design given in Table 8.2.

Latin hypercube designs are balanced on any individual factor, but are not balanced across pairs or larger sets of factors. An improvement of balance may be obtained by using as the basic design an orthogonal array of strength two, Latin hypercubes being orthogonal arrays of strength 1; see Section 6.3.

In addition to exploring the shape of an unknown and possibly highly nonlinear response surface, $y = f(x_1, \ldots, x_k)$, space-filling designs can be used to evaluate the integral

$$\int f(x) dx \quad (7.42)$$

where $f(\cdot)$ may represent a density for a k-dimensional variable x or may be the expected value of a given function $f(X)$ with respect to the uniform density. The resultant approximation to (7.42) often has smaller variance than that obtained by Monte Carlo sampling.

7.8 Bayesian design

We have in the main part of the book deliberately emphasized the essentially qualitative good properties of designs. While the methods of analysis have been predominantly based on an appropriate linear model estimated by the method of least squares, or implicitly by an analogous generalized linear model, were some other

BAYESIAN DESIGN 183

type of analysis to be used or a special *ad hoc* model developed, the designs would in most cases retain considerable appeal. In the present chapter more formal optimum properties, largely based on least squares analyses, have been given more emphasis. In many aspects of design as yet unknown features of the system under study are relevant. If the uncertainty about these features can be captured in a prior probability distribution, Bayesian considerations can be invoked.

In fact there are several rather different ways in which a Bayesian view of design might be formulated and there is indeed a quite extensive literature. We outline several possibilities.

First choosing a design is clearly a decision. In an inferential process we may conclude that there are several equally successful interpretations. In a terminal decision process even if there are several essentially equally appealing possibilities just one has to be chosen. This raises the possibility of a decision-analytic formulation of design choice as optimizing an expected utility. In particular a full Bayesian analysis would require a utility function for the various designs as a function of unknown features of the system and a prior probability distribution for those features.

Secondly there is the possibility that the whole of the objective of the study, not just the choice of design, can be formulated as a decision problem. While of course the objectives of a study and the possible consequences of various possible outcomes have always to be considered, in most of the applications we have in mind in this book a full decision analysis is not likely to be feasible and we shall not address this aspect.

Thirdly there may be prior information about the uncontrolled variation. Use of this at a qualitative level has been the primary theme of Chapters 3 and 4. The special spatial and temporal models of uncontrolled variation to be discussed in Sections 8.4 and 8.5 are also of this type although a fully Bayesian interpretation of them would require a hyperprior over the defining parameters.

Finally there may be prior information about the contrasts of primary interest, i.e. a prior distribution for the parameter of interest.

We stress that the issue is not whether prior information of these various kinds should be used, but rather whether it can be used quantitatively via a prior probability distribution and possibly a utility function.

If the prior distribution is based on explicit empirical data or theoretical calculation it will usually be sensible to use that same information in the analysis of the data. In the other, and perhaps more common, situation where the prior is rather more impressionistic and specific to experience of the individuals designing the study, the analysis may best not use that prior. Indeed one of the main themes of the earlier chapters is the construction of designs that will lead to broadly acceptable conclusions. If one insisted on a Bayesian formulation, which we do not, then the conclusions should be convincing to individuals with a broad range of priors not only about the parameters of interest but also about the structure of the uncontrolled variation.

Consider an experiment to be analysed via a parametric model defined by unknown parameters θ. Denote a possible design by ξ and the resulting vector of responses by Y. Let the prior density of θ to be used in designing the experiment be $p_{0d}(\theta)$ and the prior density to be used in analysis be $p_{0a}(\theta)$. One multipurpose utility function that might be used is the Shannon information in the posterior distribution. A more specifically statistical version would be some measure of the size of the posterior covariance matrix of θ.

Now for linear models the posterior covariance matrix under normality of both model and prior is proportional to

$$\{nM(\xi) + P_0\}^{-1}, \tag{7.43}$$

where P_0 is a contribution proportional to the prior concentration matrix of θ. Because the prior distribution in this formulation does not depend on either the responses Y or the value of θ all the non-Bayesian criteria can be used with relatively minor modification. Thus maximization of

$$\log \det\{M(\xi) + P_0/n\}$$

gives Bayesian D–optimality. Unless the prior information is strong the Bayesian formulation does not make a radical difference. Note that P_0 would typically be given by the prior to be used for analysis, should that be known. That is to say, even if hypothetically the individuals designing the experiment knew the correct value of θ but were not allowed to use that information in analysis the choice of design would be unaffected.

The situation is quite different in nonlinear problems where the optimal design typically depends on θ. Here at least in principle the preposterior expected utility is the basis for design choice. Thus with a scalar unknown parameter and squared error loss as the criterion the objective would be to minimize the expectation, taken over the prior distribution $p_{0d}(\theta)$ and over Y, of the posterior mean of θ evaluated under $p_{0a}(\theta)$. In the extreme case where $p_{0d}(\theta)$ is highly concentrated around θ_0 but the prior $p_{0a}(\theta)$ is very dispersed this amounts to using the design locally optimum at θ_0 with the non-Bayesian analysis. This design would also be suitable if interest is focused on the sign of $\theta - \theta_0$. In the notionally complementary case where the prior for analysis is highly concentrated but not to be used in design it may be a waste of resources to do an experiment at all!

It can be shown theoretically and is clear on qualitative grounds that, especially if the design prior is quite dispersed, a design with many points of support will be required.

Implementation of the above procedure to deduce an optimal design will often require extensive computation even in simple problems. One much simpler approach is to find the Fisher information, $I(\theta, \xi)$ for a given design, i.e. the expected information averaged over the responses at a fixed θ, and to maximize that averaged over the prior distribution $p_{0d}(\theta)$ of θ or slightly more generally to maximize

$$\int \phi\{I(\theta, \xi)\} p_{0d}(\theta) d\theta; \qquad (7.44)$$

for some suitable ϕ. This would not preclude the use of $p_{0a}(\theta)$ in analysis, it being assumed that this would have only a second-order effect on the choice of design.

Thus in the dilution series a single observation at dilution k contributes Fisher information

$$I(\mu, k) = k^{-2} e^{-\mu/k} (1 - e^{-\mu/k})^{-1}$$

so that the objective is to maximize

$$\int I(\mu, k) p_{0d}(\mu) d\mu \xi(dk) \qquad (7.45)$$

with respect to the design measure ξ. Note that the prior density p_{0d} must be proper. For an improper prior, evaluated as a limit of a uniform distribution of μ over $(0, A)$ as A tends to infinity, the average information at any k is zero.

This optimization problem must be done numerically, for example by computing (7.45) for discrete designs ξ with m design points and weight w_i assigned to dilution $k_i, i = 1, \ldots m$. Given a prior range for $\mu = (\mu_L, \mu_U)$, say, we will have $k_i \in (1.594/\mu_U, 1.594/\mu_L)$ as in Section 7.6. Numerical calculation based on a uniform prior for $\log \mu$ indicates that for a sufficiently narrow prior the optimal design has just one support point, but a more diffuse prior permits a larger number of support points; see the Bibliographic notes.

7.9 Optimality of traditional designs

In the simpler highly balanced designs strong optimality properties can be deduced from the following considerations.

Suppose first that interest is focused on a particular treatment contrast, not necessarily a main effect. Now if it could be assumed that all other effects are absent, the problem is essentially that of comparing two or more groups. Provided that the error variance is constant and the errors uncorrelated this is most effectively achieved by equal replication. In the case of comparison of two treatment groups this means that the variance of the comparison is that of the difference between two means of independent samples of equal size.

Next note that this variance is achieved in the standard factorial or fractional factorial systems, provided in the latter case the contrast can be estimated free of damaging aliasing.

Finally the inclusion of additional terms into a linear model cannot decrease the variance of the estimates of the parameters initially present. Indeed it will increase the variance unless orthogonality holds; see Appendix A.2.

Proof of the optimality of more complex designs, such as balanced incomplete block designs, is in most cases more difficult; see the Bibliographic notes.

7.10 Bibliographic notes

Optimal design for fitting polynomials was considered in great detail by Smith (1918). The discussion of nonlinear design for the dilution series is due to Fisher (1935). While there are other isolated investigations the first general discussions are due to Elfving (1952, 1959), Chernoff (1953) and Box and Lucas (1959). An explosion of the subject followed the discovery of the General Equivalence Theorem by Kiefer and Wolfowitz (1959). Key papers are by Kiefer (1958, 1959, 1975); see also his collected works (Kiefer, 1985). Fedorov's book (Fedorov, 1972) emphasizes algorithms for design construction, whereas the books by Silvey (1980) and Pukelsheim (1993) stress respectively briefly and in detail the underlying general theory in the linear case. Atkinson and Donev (1992) in their less mathematical discussion give many examples of optimal designs for specific situations. Important results on the convergence of algorithms like the simple one of Section 7.5 are due to Wynn (1970); see also Fedorov and Hackl (1997). There is a large literature on the use of so-called "alphabet" optimality criteria for a number of more specialized applications appearing in the theoretical statistical journals. Flournoy, Rosenberger and Wong (1998) presents recent work on nonlinear optimal design and designs achieving multiple objectives.

A lucid account of exact optimality is given by Shah and Sinha (1989). A key technique is that many of the optimality criteria are expressed in terms of the eigenvalues of the matrix C of Section 4.2. It can then be shown that unnecessary lack of balance in the design leads to sub-optimal designs by all these criteria. In that way the optimality of balanced incomplete designs and group divisible designs can be established. For the optimality of orthogonal arrays and fractional factorial designs, see Dey and Mukerjee (1999, Chapter 2).

A systematic review of Bayesian work on design of experiments is given by Chaloner and Verdinelli (1995). For the use of previous data, see Covey-Crump and Silvey (1970). A general treatment in terms of Shannon information is due to Lindley (1956). Atkinson and Donev (1992, Chapter 19) give some interesting examples. Dawid and Sebastiani (1999) have given a general discussion treating design as part of a fully formulated decision problem.

Bayesian design for the dilution series is discussed in Mehrabi and Matthews (1998), where in particular they illustrate the use of diffuse priors to obtain a four-point design and more concentrated priors to obtain a one-point design. General results on the relation of the prior to the number of support points are described in Chaloner (1993).

Box and Draper (1959) emphasized the importance of space-filling designs when the goal is prediction of response at a new design point, and when random variability is of much less importance than systematic variability. Latin hypercube designs were proposed in McKay, Beckman and Conover (1979). Sacks, Welch, Mitchell and Wynn (1989) survey the use of optimal design theory for computer experiments; i.e. simulation experiments in which runs are expensive and the output is deterministic. Aslett et al. (1998) give a detailed case study of a sequential approach to the design of a circuit simulator, where a Latin hypercube design is used in the initial stages. For applications to models of BSE and vCJD, see Donnelly and Ferguson (1999, Chapters 9, 10). Owen (1993) proves a central limit theorem for Latin hypercube sampling, and indicates how these samples may be used to explore the shape of the response function, for example in finding the maximum of f over the design space. He considers the more general case of evaluating the expected value of f with respect to a known density g.

Another approach to space-filling design using methods from number theory is briefly described in Exercise 7.7. This approach is reviewed by Fang, Wang and Bentler (1994) and its application in design of experiments discussed in Chapter 5 of Fang and Wang (1993). In the computer science literature the method is often called quasi-Monte Carlo sampling; see Neiderreiter (1992).

7.11 Further results and exercises

1. Suppose that an optimal design for a model with p parameters has support on more than $p(p+1)/2$ distinct points. Then because an arbitrary information matrix can be formed from a convex combination of those for $p(p+1)/2$ points there must be an optimal design with only that number of points of support. Often fewer points, indeed often only p points, are needed.

2. Show that a number of types of optimality criterion can be encapsulated into a single form via the eigenvalues $\gamma_1, \ldots, \gamma_p$ of the matrix $M(\xi)$ on noting that the eigenvalues of $M^{-1}(\xi)$ are the reciprocals of the γ_s and defining
$$\Pi_k(\xi) = (p^{-1}\Sigma\gamma_s^{-k})^{1/k}.$$
Examine the special cases $k = \infty, 1, 0$. See Kiefer (1975).

3. Optimal designs for estimating logistic regression allowing for robustness to parameter choice and considering sequential possibilities were studied by Abdelbasit and Plackett (1983). Heise and Myers (1996) introduced a bivariate logistic model, i.e. with two responses, for instance efficacy and toxicity. They defined Q optimality to be the minimization of an average over the design region of a variance of prediction and developed designs for estimating the probability of efficacy without toxicity. See several

papers in the conference volume edited by Flournoy et al. (1998) for further extensions.

4. Physical angular correlation studies involve placing pairs of detectors to record the simultaneous emission of pairs of particles whose paths subtend an angle θ at a source. There are theoretical grounds for expecting the rate of emission to be given by a few terms of an expansion in Legendre polynomials, namely to be of the form

$$\beta_0 + \beta_2 P_2(\cos\theta) + \beta_4 P_4(\cos\theta) =$$
$$\beta_0 + \beta_2(3\cos^2\theta - 1)/2 + \beta_4(35\cos^4\theta - 30\cos^2\theta + 3)/8.$$

Show, assuming that the estimated counting rate at each angle is approximately normal with constant variance, that the optimal design is to choose three equally spaced values of $\cos^2\theta$, namely values of θ of 90, 135 and 180 degrees. If the resulting counts have Poisson distributions, with largish means, what are the implications for the observational times at the three angles?

5. In an experiment on the properties of materials cast from high purity metals it was possible to cast four 2 kg ingots from a 8 kg melt. After the first ingot from a melt had been cast it was possible to change the composition of the remaining molten metal but only by the *addition* of one of the alloying metals, no technique being available for selectively reducing the concentration of an alloying metal. Thus within any one melt and with one alloying metal at two levels $(1, a)$ there are five possible sequences namely

$$(1,1,1,1); (1,1,1,a); (1,1,a,a); (1,a,a,a); (a,a,a,a),$$

with corresponding restrictions on each factor separately if a factorial experiment is considered. This is rather an extreme example of practical constraints on the sequence of experimentation.

By symmetry an optimal design is likely to have

$$n_1, n_2, n_3, n_2, n_1$$

melts of the five types. Show that under the usual model allowing for melt (block) and period effects the information about the treatment effect is proportional to

$$12n_1 n_2 + 8n_1 n_3 + 6n_2 n_3 + 4n_2^2.$$

By maximizing this subject to the constraint that the total number of melts $2n_1+2n_2+n_3$ is fixed show that the simplest realizable optimal design has 8 melts with $n_1 = 1, n_2 = n_3 = 2$. Show that this is an advantageous design as compared with holding the treatment fixed within a melt, i.e. taking a whole melt as an experimental unit if and only if there is a substantial component of variance between melts. For more details including the factorial case, see Cox (1954).

6. In Latin hypercube sampling, if we sample randomly for each subinterval, we can represent the resulting sample $(X_1,\ldots,X_n)$, or in component form $(X_{11},\ldots,X_{1m};\ldots;X_{n1},\ldots,X_{nm})$, by

$$X_{ij} = (\pi_{ij} - U_{ij})/n, \quad i = 1,\ldots n, j = 1,\ldots,m, \quad (7.46)$$

where $\pi_{1j},\ldots,\pi_{nj}$ is a random permutation of $\{1,\ldots,n\}$ and U_{ij} is a random variable distributed uniformly on $(0,1)$. Suppose our goal is to evaluate $Y = f(X)$, $X \in R^m$ and $Y \in R$ where f is a known function but very expensive to compute. Show that $\text{var}(Y)$ under the sampling scheme (7.46) is

$$n^{-1}\text{var}\{f(X)\} - n^{-1}\sum_{j=1}^{m}\text{var}\{f_j(X_j)\} + o(n^{-1}), \quad (7.47)$$

where $f_j(X_j) = E\{f(X)|X_j\} - E\{f(X)\}$ is the "main effect" of $f(X)$ for the jth component of X. For details see Stein (1987) and Owen (1992). Tang (1993) shows how to reduce the variance further by orthogonal array based sampling.

7. Another type of space-filling design specifies points in the design space using methods from number theory. The resulting design is called a *uniform*, or *uniformly scattered* design. In one dimension a uniformly scattered design of n points is simply

$$\{(2i-1)/(2n), i = 1,\ldots,n\}. \quad (7.48)$$

This design has the property that it minimizes, among all n-point designs, the Kolmogorov-Smirnov distance between the design measure and the uniform measure on $(0,1)$:

$$D_n(\xi) = \sup_{x \in (0,1)} |F_n(x) - x| \quad (7.49)$$

where $F_n(x) = n^{-1}\sum 1\{\xi_i \leq x\}$ is the empirical distribution function for the design $\xi = (\xi_1,\ldots\xi_n)$.

In $k \geq 2$ dimensions the determination of a uniformly scattered design, i.e. a design that minimizes the Kolmogorov-Smirnov distance between the design measure and the uniform measure on $[0,1]^k$ is rather difficult and is often simplified by seeking designs that achieve this property asymptotically. Detailed results and a variety of applications are given in Fang and Wang, (1993, Chapters 1, 5).

CHAPTER 8

Some additional topics

8.1 Scale of effort

8.1.1 General remarks

An important part of the design of any investigation involves the scale of effort appropriate. In the terminology used in this book the total number of experimental units must be chosen. Also, commonly, a decision must be made about the number of repeat observations to be made on important variables within each unit, especially when sampling of material within each unit is needed.

Illustration. In measuring yield of product per plot in an agricultural field trial the whole plot might be harvested and the total yield measured; in some contexts sample areas within the plot would be used. In measuring other aspects concerned with the quality of product, sampling within each plot would be essential. Similar remarks apply to the product of, for example, chemical reactions, when chemical analyses would be carried out on small subsamples.

It is, of course, crucial to distinguish between the number of experimental units and the total number of observations, which will be much greater if there is intensive sampling within units. Precision of treatment contrasts is usually determined primarily by the number of units and only secondarily by the number of repeat observations per unit.

Often the investigator has appreciable control over the amount of sampling per unit. As regards the number of units, there are two broad situations.

In the first the number of units is largely outside the investigator's control. The question at issue is then usually whether the resources are adequate to produce an answer of sufficient precision to be useful and hence to justify proceeding with the investigation. Just occasionally it may be that the resources are unnecessarily great and that it may be sensible to begin with a smaller investigation. The second possibility is that there is a substantial degree of control over the number of units to use and in that case some calculation of the number needed to achieve reasonable precision is required. In both cases some initial consideration of the number of units is very desirable although, as we shall see, there is substantial arbitrariness involved and elaborate calculations of high apparent precision are very rarely if ever justified.

If the investigation is of a new or especially complex kind it will be important to do a pilot study of as many aspects of the procedure as feasible. The resources devoted to this will, however, typically be small compared with the total available and this aspect will not be considered further.

A further general aspect concerns the time scale of the investigation. If results are obtained quickly it may be sensible to proceed in relatively small steps, calculating confidence limits from time to time and stopping when adequate precision has been achieved. To avoid possible biases it will, however, be desirable to set a target precision in advance.

A decision-theoretic formulation is outlined in a later subsection, but we deal first with some rather informal procedures that are commonly used. The importance of these arguments in fields such as clinical trials, where they are widely applied, is probably in ensuring some uniformity of procedure. If experience indicates that a certain level of precision produces effective results the calculations provide some check that in each new situation broadly appropriate procedures are being followed and this, of course, is by no means the same as using the same number of experimental units in all contexts.

We deal first with the number of experimental units and subsequently with the issue of sampling within units. In much of the discussion we consider an experiment to compare two treatments, T and C, the extension to more than two treatments and to simple factorial systems being immediate.

8.1.2 Precision and power

We have in the main discussion in this book emphasized the objective of estimating treatment contrasts, in the simplest case differences between pairs of treatments and in more complex cases factorial contrasts. In the simplest cases these are estimated with a standard error of the form $\sigma\sqrt{(2/m)}$, where m is related to the

SCALE OF EFFORT   195

total number n of experimental units. For example, if comparison of pairs of treatments is involved, $m = n/v$, when v equally replicated treatments are used. The standard deviation σ is residual to any blocking system used. Approximate confidence limits are directly calculated from the standard error.

A very direct and appealing formulation is to aim that the standard error of contrasts of interest is near to some target level, d, say. This could also be formulated in terms of the width of confidence intervals at some chosen confidence levels. Direct use of the standard error leads to the choice

$$m = 2\sigma^2/d^2 \tag{8.1}$$

and hence to an appropriate n. If the formulation is in terms of a standard error required to be a given fraction of an overall mean, for example 0.05 times the overall mean, σ is to be interpreted as a coefficient of variation. If the comparison is of the means of Poisson variables or of the probabilities of binary events essentially minor changes are needed.

Now while we have put some emphasis in the book on the estimation of precision internally from the experiment itself, usually via an appropriate residual sum of squares, it will be rare that there is not some rough idea from previous experience about the value of σ^2. Indeed if there is no such value it will probably be particularly unwise to proceed much further without a pilot study which, in particular, will give an approximate value for σ^2. Primarily, therefore, we regard the above simple formula as the one to use in discussions of appropriate sample size. Note that to produce substantial improvements in precision via increased replication large changes in m and n are needed, a four-fold increase to halve the standard error. We return to this point below.

A conceptually more complicated but essentially equivalent procedure is based on the power of a test of the null hypothesis that a treatment contrast, for example a simple difference, is zero. If Δ_0 represents a difference which we wish to be reasonably confident of detecting if present we may require power $(1 - \beta)$ in a test at significance level α to be achieved at $\Delta = \Delta_0$. If we consider a one-sided test for simplicity this leads under normality to

$$\Delta_0 = (k_\alpha + k_\beta)\sigma\sqrt{(2/m)}, \tag{8.2}$$

where $\Phi(-k_\alpha) = \alpha$.

This is equivalent to requiring the standard error to be $\Delta_0/(k_\alpha + k_\beta)$.

Quite apart from the general undesirability of focusing the objectives of experimentation on hypothesis testing, note that the interpretation in terms of power requires the specification of three somewhat arbitrary quantities many of them leading to the same choice of m. For that reason a formulation directly in terms of a target standard error is to be preferred. A further general comment concerns the dependence of standard error on m or n. The inverse square-root dependence holds so long as the effective standard deviation, σ, does not depend on n. In practice, for a variety of reasons, if n varies over a very wide range it is quite likely that σ increases slowly with n, for example because it is more difficult to maintain control over large studies than over small studies. For that reason the gains in precision achieved by massive increases in n are likely to be less than those predicted above.

Finally, if observations can be obtained and analysed quickly it may be feasible simply to continue an investigation until the required standard error, d, has been achieved rather than having a prior commitment to a particular n.

8.1.3 Sampling within units

We now suppose that on each of n experimental units r repeat observations are taken, typically representing a random sample of material forming the unit. Ignoring the possibility that the sampled material is an appreciable proportion of the whole, so that a finite population correction is unnecessary, the effective variance per unit is

$$\sigma_b^2 + \sigma_w^2/r, \tag{8.3}$$

where σ_b^2 and σ_w^2 are respectively components of variance between and within units: see Appendix A.3. The precision of treatment comparisons is determined by

$$(\sigma_b^2 + \sigma_w^2/r)/n \tag{8.4}$$

and we assume that the cost of experimentation is proportional to

$$\kappa n + rn, \tag{8.5}$$

where κ is the ratio of the cost of a unit to the cost of a sampled observation within a unit.

We may now either minimize the variance for a given cost or minimize the cost for a given variance, which essentially leads to minimizing the objective function

$$(\sigma_b^2 + \sigma_w^2/r)(\kappa + r). \tag{8.6}$$

The optimum value of r for the estimation of treatment contrasts is $\kappa^{1/2}\sigma_w/\sigma_b$. Also, because the third derivative of the objective function is negative, the function falls relatively steeply to and rises relatively slowly from its minimum so that it will often be best to take the integer larger than the in general non-integer value given by the formula. The arguments for this are accentuated if either estimation of the components of variance is of intrinsic interest or if it is required to take special precautions against occasional bad values.

There will usually be strong arguments for taking the same value of r for all units. A possible exception is when the formal optimum given above is only slightly greater than one when a balanced subsample of units should be selected, preferably with an element of randomization, to be sampled twice.

A final general point is that observations on repeat samples within the same unit should be measured blind to their identity. Otherwise substantial underestimation of error may arise and at least some of the advantages of internal replication lost.

8.1.4 Quasi-economic analysis

In principle the choice of number of units involves a balance between the cost of experimental units and the losses arising from imprecision in the conclusions. If these can be expressed in common units, for example of time or money, an optimum can be determined. This is a decision-oriented formulation and involves formalizing the objective of the investigation in decision-theoretic terms.

While, even then, it may be rather rare to be able to attach meaningful numbers to the costs involved it is instructive to examine the resulting formulae. There are two rather different cases depending on whether an essentially continuous choice or a discrete one is involved in the experiment.

As probably the simplest example of the first situation suppose that a regression relation of the form

$$E(Y; x) = \beta_0 + \beta_1 x + \beta_2 x^2 = m(x), \tag{8.7}$$

is investigated, for $x \in [-1, 1]$, with a design putting $n/3$ observations at each of $x = -1, 0, 1$. Assume that $\beta_2 > 0$ so that the minimum response is at $\theta = -\beta_1/(2\beta_2)$ and that this is estimated via the least squares estimates of β_1, β_2 leading to $\hat{\theta} = -\hat{\beta}_1/(2\hat{\beta}_2)$.

Now suppose that Y represents the cost per unit yield, and the objective is to achieve minimum response in future applications. The loss compared with complete knowledge of the parameters can thus be measured via

$$m(\hat{\theta}) - m(\theta) = \beta_1(\hat{\theta} - \theta) + \beta_2(\hat{\theta}^2 - \theta^2), \tag{8.8}$$

with expected value

$$\beta_2 E(\hat{\theta}^2) + \beta_1 E(\hat{\theta}) + \frac{\beta_1^2}{4\beta_2}. \tag{8.9}$$

Now make the optimistic assumption that via preliminary investigation the design has been correctly centred so that, while β_1 has still to be estimated its true value is small and that curvature is the predominant effect in the regression. Then approximately

$$\text{var}(\hat{\theta}) = \text{var}(\hat{\beta}_1)/(4\beta_2^2) \tag{8.10}$$
$$= 3\sigma^2/(8n\beta_2^2) \tag{8.11}$$

under the usual assumptions on error.

If now the conclusions are applied to a target population equivalent to N experimental units and if c_y is the cost of unit increase in Y per amount of material equivalent to one experimental unit, then the expected loss arising from errors in estimating θ is

$$\frac{3Nc_y\sigma^2}{8n\beta_2}, \tag{8.12}$$

whereas the cost of experimentation, ignoring set up cost and assuming the cost c_n per unit is constant, is nc_n leading to an optimum value of the number of units of

$$\left(\frac{3Nc_y\sigma^2}{8c_n\beta_2}\right)^{1/2}. \tag{8.13}$$

This has the general dependence on the defining parameters that might have been anticipated; note, however, that the approximations involved preclude the use of the formula in the "flat" case when both β_1, β_2 are very small.

Now suppose that a choice between just two treatments T, C is involved. We continue to suppose that the response variable Y is such that its value is to be minimized and assume that for application to a target population equivalent to N units the difference in costs is Nc_y times the difference of expected values under the two treatments, i.e. is $Nc_y\Delta$, say. On the basis of an experiment involving n experimental units we estimate Δ by $\hat{\Delta}$ with variance $4\sigma^2/n$. For simplicity suppose that there is no *a priori* cost difference between the treatments so that the treatment with smaller mean is chosen. Then the loss arising from errors of estimation is zero if Δ and $\hat{\Delta}$ have the same sign and is $Nc_y|\Delta|$ otherwise. For given Δ the expected loss is thus

$$Nc_y|\Delta|P(\Delta\hat{\Delta} < 0; \Delta) = Nc_y|\Delta|\Phi\{-|\Delta|\sqrt{n}/(2\sigma)\}. \qquad (8.14)$$

The total expected cost is this plus the cost of experimentation, taken to be $c_n n$.

The most satisfactory approach is now to explore the total cost as a function of n for a range of plausible values of Δ, the dependence on Nc_y/c_n being straightforward. A formally more appealing, although often ultimately less insightful, approach is to average over a prior distribution of Δ. For simplicity we take the prior to be normal of zero mean and variance τ^2: except for a constant independent of n the expected cost becomes

$$-\frac{Nc_y\tau\sqrt{n}}{\sqrt{(2\pi)}}(n + \frac{4}{\kappa^2})^{-1/2} + nc_n, \qquad (8.15)$$

where $\kappa = \tau/\sigma$. The optimum n can now be found numerically as a function of κ and of $Nc_y\tau/c_n$. If κ is very large, a situation in which a large value of n will be required, the optimum n is approximately

$$\frac{Nc_y\tau}{\sqrt{(2\pi)}c_n}. \qquad (8.16)$$

The dependence on target population size and cost ratio is more sensitive than in the optimization problem discussed above presumably because of the relatively greater sensitivity to small errors of estimation.

When the response is essentially binary, for example satisfactory and unsatisfactory, an explicit assignment of cost ratios may sometimes be evaded by supposing that the whole target population is available for experimentation, that after an initial phase in which n units are randomized between T and C the apparently superior treatment is chosen and applied to the remaining $(N-n)$ units. The objective is to maximize the number of individuals giving satisfactory response, or, equivalently, to maximize the number of individuals receiving the superior treatment. Note that $n/2$ individuals receive the inferior treatment in the experimental phase. The discussion assumes absence of a qualitative treatment by individual interaction.

We argue very approximately as follows. Suppose that the proportions satisfactory are between, say 0.2 and 0.8, for both treatments. Then the difference Δ between the proportions satisfactory is estimated with a standard error of approximately $0.9/\sqrt{n}$; this is because the binomial variance ranges from 0.25 to 0.16 over the proportions in question and 0.2 is a reasonable approximation for exploratory purposes. The probability of a wrong choice is thus $\Phi(-|\Delta|\sqrt{n}/0.9)$ and the expected number of wrongly treated individuals is

$$n/2 + (N-n)\Phi(-|\Delta|\sqrt{n}/0.9). \tag{8.17}$$

Again we may either explore this as a function of (n, Δ) or average over a prior distribution of Δ. With a normal prior of zero mean and variance τ^2 we obtain an expected number of wrongly treated individuals of

$$\frac{n}{2} + \frac{N-n}{\pi}\tan^{-1}(\frac{0.9}{\tau\sqrt{n}}). \tag{8.18}$$

For fixed N and τ the value of n minimizing (8.18), n_{opt}, say, is readily computed. The optimum proportion, n_{opt}/N is shown as a function of N and τ in Figure 8.1. This proportion depends more strongly on τ than on N, especially for τ between about 0.5 and 3. As τ approaches 0 the proportion wrongly treated becomes 0.5, and as τ approaches ∞ the number wrongly treated is $n/2$.

While the kinds of calculation outlined in this section throw some light on the considerations entering choice of scale of effort their immediate use in applications is limited by two rather different considerations. First the explicit formulation of the objective of experimentation in a decision-making framework may be inapplicable. Thus, even though the formulation of maximizing the number of correctly treated individuals is motivated by clinical trial applications, it is nearly always a considerable oversimplification

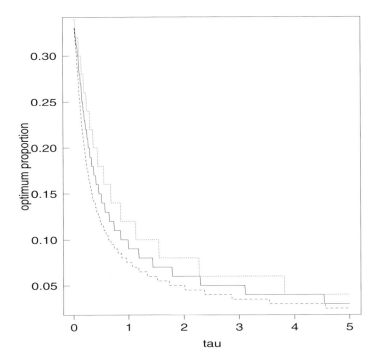

Figure 8.1 *Optimum proportion to be entered into experiment, n_{opt}/N, as a function of target population, N, and prior variance τ of the difference of proportions: $N = 100$ (———), 200 (- - -), 50 (· · · · · ·).*

to suppose that strategies for treating whole groups of patients are determined by the outcome of one study. Secondly, even if the decision-theoretic formulation of objectives is reasonably appropriate, it may be extremely hard to attach meaningful numbers to the quantities determining n.

For that reason the procedures commonly used in considerations of appropriate n are the more informal ones outlined in the earlier subsections.

8.2 Adaptive designs

8.2.1 General remarks

In the previous discussion it has been assumed implicitly that each experiment is designed and implemented in one step or at least in substantial sections dealt with in turn. If, however, experimental units are used in sequence and the response on each unit is obtained quite quickly, much greater flexibility in design is possible. We consider mostly an extreme case in which the response of each unit becomes available before the next unit has to be randomized to its treatment. Similar considerations apply in the less extreme case where experimental units are dealt with in smallish groups at a time or there is a delay before each response is obtained, during which a modest number of further experimental units can be entered. The greater flexibility can be used in various ways. First, we could choose the size of the experiment in the light of the observations obtained, using so-called *sequential stopping rules*. Secondly, we might modify the choice of experimental units, treatments and response variables in the light of experience. Finally, we could allocate a treatment to the next experimental unit in the light of the responses so far obtained.

Illustrations. An agricultural field trial will typically take a growing season to produce responses, a rotation experiment much longer and an experiment on pruning fruit trees many years. Adaptive treatment allocation is thus of little or no relevance. By contrast industrial experiments, especially on batch processes, and some laboratory experiments may generate responses very quickly once the necessary techniques are well established. Clinical trials on long-term conditions such as hypertension may involve follow-up of patients for several years; on the other hand where the immediate relief of symptoms of, for instance, asthma is involved, adaptive allocation may become appropriate.

We concentrate here largely on adaptive treatment allocation, with first some brief comments on modifications of the primary features of the experiment.

8.2.2 Modifications of primary features

Considered on a sufficiently long time-scale the great majority of experimentation is intrinsically sequential; the analysis and interpretation of one experiment virtually always raises further questions for investigation and in the meantime other relevant work may well have appeared. Thus the possibility of frequent or nearly continuous updating of design and objectives is, at first sight, very appealing, but there are dangers. Especially in fields where new

ADAPTIVE DESIGNS

ideas arise at quite high speed, there can be drawbacks to changing especially the focus of investigations until reasonably unambiguous results have been achieved.

The broad principles underlying possible changes in key aspects of the design in the course of an experiment are that it should be possible to check for a possible systematic shift in response after the change and that a unified analysis of the whole experiment should be available. For example, changes in the definition of an experimental unit may be considered.

Illustration. Suppose that in a clinical trial patients are, in particular, required to be in the age range 55–65 years, but that after some time it is found that patient accrual is much slower than expected. Then it may be reasonable to relax the trial protocol to allow a wider age range.

In general it may be proposed to change the acceptable values of one or more baseline variables, z. Analysis of the likely effects of this change is, of course, important before a decision is reached. One possibility is that the treatment effects under study interact with z. This might take the form of a linear interaction with z, or, perhaps more realistically in the illustration sketched above, the approximation that a treatment difference Δ under the original protocol becomes a treatment difference $\kappa\Delta$ for individuals who pass the new, but not the old, entry requirements; here κ is an unknown constant, with probably $0 < \kappa < 1$, especially if the original protocol was formulated to include only individuals likely to show a treatment effect in the most sensitive form. Another possibility is that while the treatment effect Δ is the same for the new individual the variance is increased.

A rather crude analysis of this situation is as follows. Suppose that an analysis is considered in which, if the protocol is extended the null hypothesis $\Delta = 0$ is tested via the mean difference over all individuals. Let there be m units on each treatment a proportion p of which meet the original protocol. The sensitivity of a test using only the individuals meeting the original protocol will be determined by the quantity

$$\Delta\sqrt{(pm)}/\sigma, \tag{8.19}$$

where σ is the standard deviation of an individual difference. For the mean difference over all individuals the corresponding measure

is
$$\{p\Delta + (1-p)\kappa\Delta\}\sqrt{m}/[\sigma\sqrt{\{p+\gamma(1-p)\}}], \qquad (8.20)$$

where $\gamma \geq 1$ is the ratio of the variance of the individuals in the extended protocol to those in the original.

There is thus a formal gain from including the new individuals if and only if

$$p + (1-p)\kappa > \{p^2 + p(1-p)\gamma\}^{1/2}. \qquad (8.21)$$

The formal advantages of extending the protocol would be greater if a weighted analysis were used to account for differences of variance possibly together with an ordered alternative test to account for the possible deflation of the treatment effect.

Note that amending the requirements for entry into an experiment is not the same as so-called *enrichment entry*. In this all units are first tested on one of the proposed treatments and only those giving appropriate responses are randomized into the main experiment. There are various possible biases inherent in this procedure.

Modification of the treatments used in the light of intermediate results is most likely by:

1. omitting treatments that appear to give uninteresting results;
2. introducing new and omitting current factors in fractional factorial experiments (see Section 5.6);
3. changing the range of levels used for factors with quantitative levels; see Section 8.3 below.

If the nature of the response variable is changed, it will be important to ensure that the change does not bias treatment contrasts but also, unless the change is very minor, to aim to collect both new and old response variable on a nontrivial number of experimental units; see Exercise 8.1.

8.2.3 Modification of treatment allocation: two treatments

We now consider for simplicity the comparison of two treatments T, C; the arguments extend fairly directly to the comparison of any small number of treatments. If the primary objective is the estimation of the treatment difference, then there will be no case for abandoning equal replication unless either the variance of the response or the cost of implementation is different for the two treatments. For example, if T involves extensive modifications or novel substances it may be appreciably more expensive than C and this may only become apparent as the work progresses. If, however, an objective is, as in Section 8.1, to treat as many individuals as possible successfully, there are theoretically many possible ways of implementing adaptive allocation, either to achieve economy or to address ethical considerations. We consider in more detail the situation of Section 8.1 with binary responses.

The formulation used earlier involved a target population of N individuals, randomized with equal representation of T and C, until some point after which all individuals received the same treatment. Clearly there are many possibilities for a smoother transition between the two regimes which, under some circumstances may be preferable. The question of which is in some sense optimum depends rather delicately on the balance between the somewhat decision-like formulation of treating individuals optimally and the objective of estimating a treatment difference and may also involve ethical considerations.

The *play-the-winner* rule in its simplest form specifies that if and only if the treatment applied to the rth unit yields a success the same treatment is used for the $(r+1)$st unit. Then if this continues for a long time the proportion of individuals allocated to T tends to $\theta_C/(\theta_T+\theta_C)$, where θ_T and θ_C are respectively the probabilities of failure under T and C. Unless the ratio of these probabilities is appreciable the concentration of resources on the superior treatment is thus relatively modest. Also, in some contexts, the property that the treatment to be assigned to a new unit is known as soon as the response on the previous unit is available would be a source of serious potential bias. There are various modifications of the procedure incorporating some randomization that would overcome this to some extent.

A further point is that if after n units have been used the numbers of units and numbers of successes are respectively n_T, n_C and r_T, r_C under T and C, then these are sufficient statistics for the defining parameters provided these parameters are stable in time. A concept of what might be called *design sufficiency* suggests that if a data-dependent allocation rule is to be used it should be via the sufficient statistics. The play-the-winner rule is a local one and is not in accord with this. This suggests that it can be improved unless the success probabilities vary in time, in which case a more local rule might indeed be appealing.

There are many possibilities for an allocation rule in which the probability that the $(r+1)$st unit is allocated to T depends on the current sufficient statistic. The simplest is the *biased coin* randomization scheme in which the probability of allocation to T is $1/2 + c$ if the majority of past results favour T, $1/2 - c$ if the majority favour C and $1/2$ if there is balance; in some versions the probability of $1/2$ is maintained for the first n_0 trials. Here c is a positive constant, equal say to $1/6$, so that the allocation probabilities are $2/3$, $1/3$ and $1/2$, respectively.

Much more generally, in any experiment in which units enter and are allocated to treatments in order, biased coin randomization can be used to steer the design in any desired direction, for example towards balance with respect to a set of baseline variables, while retaining a strong element of randomization which might be lost if exact balance were enforced.

The simplest illustration of this idea is to move towards equal replication. Define an indicator variable

$$\Delta_r = \begin{cases} 1 & \text{if } r\text{th unit receives } T \\ -1 & \text{if } r\text{th unit receives } C \end{cases} \qquad (8.22)$$

and write $S_r = \Delta_1 + \ldots + \Delta_r$. Then biased coin randomization can be used to move towards balance by making the probability that $\Delta_{r+1} = 1$ equal to $1/2 - c, 1/2, 1/2 + c$ according as $S_r >, =, < 0$.

While it is by no means essential to base the analysis of such a design on randomization theory some broad correspondence with that theory is a good thing. In particular the design forces fairly close balance in the treatment allocation over medium-sized time periods. Thus if there were a long term trend in the properties of the experimental units that trend would largely be eliminated and thus an analysis treating the data as two independent samples would overestimate error, possibly seriously.

Some of the further arguments involved can be seen from a simple special case. This is illustrated in Table 8.1, the second line of which, for example, has probability

$$1/2 \times 1/3 \times 1/3 \times 2/3. \qquad (8.23)$$

To test the null hypothesis of treatment equivalence, i.e. the null hypothesis that the observation on any unit is unaffected by treatment allocation, a direct approach is first to choose a suitable test statistic, for example the difference in mean responses between the two treatments. Then under the null hypothesis the value of the statistic can be computed for all possible treatment configurations,

ADAPTIVE DESIGNS

the probability of each evaluated and hence the null hypothesis distribution of the test statistic found. Unfortunately this argument is inadequate. For example, in two of the arrangements of Table 8.1, all units receive the same treatment and hence provide no information about the hypothesis. More generally the more balanced the arrangement the more informative the data. This suggests that the randomization should be conditioned on a measure of the balance of the design, for example on the terminal value of S_r. If a time trend in response is suspected the randomization could be conditioned also on further properties concerned with any time trend in treatment balance; of course this requires a trial of sufficient length.

This makes the randomization analysis less simple than it might appear.

To study the consequences analytically we amend the randomization scheme to one in which for some suitable small positive value of ϵ we arrange that

$$E(\Delta_{r+1} \mid S_r = s) = -\epsilon s. \tag{8.24}$$

For small ϵ the S_r form a first order autoregressive process. If we define the test statistic T_n to be the difference $\Sigma y_r \Delta_r$, where y_r is the response on the rth unit, it can be shown that asymptotically (T_n, S_n) has a bivariate normal distribution of zero mean and

Table 8.1 *Treatment allocation for four units; biased coin design with probabilities* 1/3, 1/2, 2/3. *There are 8 symmetrical arrangements starting with* −1.

Treat alloc				S_4	Prob
1	1	1	1	4	1/54
1	1	1	−1	2	2/54
1	1	−1	1	2	2/54
1	−1	1	1	2	3/54
1	1	−1	−1	0	4/54
1	−1	1	−1	0	6/54
1	−1	−1	1	0	6/54
1	−1	−1	−1	−2	3/54

covariance matrix specified by

$$\text{var}(T_n) = nc_0 - n\epsilon\Sigma c_r(1-\epsilon)^r, \qquad (8.25)$$
$$\text{cov}(S_n, T_n) = -\epsilon\Sigma_{r\neq s} y_r(1-\epsilon)^{|r-s|}/2, \qquad (8.26)$$
$$\text{var}(S_n) = (2\epsilon)^{-1}, \qquad (8.27)$$

where

$$c_k = n^{-1}\Sigma Y_r Y_{r+k} \qquad (8.28)$$

and the sum in the first formula is from $r = 1, \ldots, n-1$.

This suggests that for a formal asymptotic theory one should take ϵ to be proportional to $1/n$.

An approximate test can now be constructed from the asymptotically normal conditional distribution of T_n given $S_n = s$ which is easily calculated from the covariance matrix. More detailed calculation shows that if there is long- or medium-term variation in the y_r then the variance of the test statistic is much less than that corresponding to random sampling.

In this way some protection against bias and sometimes a substantial improvement over total randomization are obtained. The main simpler competitor is a design in which exact balance is enforced within each block of $2k$ units for some suitable fairly small k. The disadvantage of this is that after some point within each block the allocation of the next unit may be predetermined with a consequent possibility of selection bias.

8.3 Sequential regression design

We now consider briefly some of the possibilities of sequential treatment allocation when each treatment corresponds to the values of one or more quantitative variables. Optimal design for various criteria within a linear model setting typically leads to unique designs, there being no special advantage in sequential development unless the model to be fitted or the design region change. Interesting possibilities for sequential design arise either with nonlinear models, when a formally optimal design depends on the unknown true parameter value, or with unusual objectives.

As an example of the second, we consider a simple regression problem in which the response Y depends on a single explanatory variable x, so that $E(Y; x) = \psi(x)$, where $\psi(x)$ may either be an unknown but typically monotonic function or in some cases may be parameterized. Suppose that we wish to estimate the value of

x_p, say, assumed unique, such that

$$\psi(x_p) = p. \tag{8.29}$$

A special case concerns binary responses when x_p is the factor level at which a proportion p of successful responses is achieved. If $p = 1/2$ this is the so-called ED50 point when the factor level is dose of a drug.

There are two broad classes of procedure which can now be used for sequential allocation. In one there is a preset collection of levels of x, often equally spaced, and a rule is specified for moving between them. In the more elaborate versions the set of levels of x varies, typically shrinking as the target value is neared.

Note that if the function $\psi(x)$ is parameterized, optimal design for the appropriate parametric function can be used and this will typically require dispersed values of x. The methods outlined here are primarily suitable when $\psi(x)$ is to be regarded nonparametrically.

For a fixed set of levels, in the so-called up and down or staircase method a rule is specified for moving to the next higher or next lower level depending usually only on the current observation. Note that this will conflict with design sufficiency. The rule is chosen so that the levels of x used cluster around the target; estimation is typically based on an average of the levels used eliminating a transient effect arising from the initial conditions. Thus with a continuous response the simplest rule, when $\psi(x)$ is increasing, is to move up if the response is less than p, down if it is greater than p. If the rule depends only on the last response observed the system forms a Markov chain.

The most widely studied version in which the levels of x vary as the experiment proceeds is the Robbins-Monro procedure in which

$$x_{r+1} = x_r + ar^{-1}(x_r - p), \tag{8.30}$$

where $a > 0$ is a constant to be chosen. This is in effect an up and down method in which the step length decreases at a rate proportional to $1/n$; see Exercise 8.2.

8.4 Designs for one-dimensional error structure

We now turn to some designs based on quite strong assumptions about the form of the uncontrolled variation; for some preliminary remarks see Section 3.7. We deal with experimental units arranged in sequence in time or along a line in space.

Suppose then that the experimental units are arranged at equally spaced intervals in one dimension. The uncontrolled variation thus in effect defines a time series and two broad possibilities are first that there is a trend, or possibly systematic seasonal variation with known wavelength, and secondly that the uncontrolled variation shows serial correlation corresponding to a stationary time series.

In both cases, especially with fairly small numbers of treatments, grouping of adjacent units into blocks followed by use of a randomized block design or possibly a balanced incomplete block design will often supply effective error control without undue assumptions about the nature of the error process. Occasionally, however, especially in a very small experiment, it may be preferable to base the design and analysis on an assumed stochastic model for the uncontrolled variation. This is particularly likely to be useful if there is some intrinsic interest in the structure of the uncontrolled variation.

Illustration. In a textile experiment nine batches of material were to be processed in sequence comparing three treatments, T_1, T_2, T_3 equally replicated. The batches were formed by thoroughly mixing a large consignment of raw material and dividing it at random into nine equal sections, numbered at random 1, ..., 9 and to be processed in that order. The whole experiment took appreciable time and there was some possibility that the oil on the material would slowly oxidize and induce a time trend in the responses on top of random variation; there was some intrinsic interest in such a trend. It is thus plausible to assume that in addition to any treatment effect there is a time trend plus random variation.

Motivated by such situations suppose that the response on the sth unit has the form

$$Y_s = \mu + \tau_{[s]} + \beta_1 \phi_1(s) + \ldots + \beta_p \phi_p(s) + \epsilon_s, \qquad (8.31)$$

where $\tau_{[s]}$ is the treatment effect for the treatment applied to unit s and $\phi_q(s)$ is the value at point s of the orthogonal polynomial of degree q defined on the design points and the ϵ's are uncorrelated errors of zero mean and variance σ^2. Quite often one would take $p = 1$ or 2 corresponding to a linear or quadratic trend.

For example with $n = 8$ the values of $\phi_1(s), \phi_2(s)$ are respectively

DESIGNS FOR ONE-DIMENSIONAL ERROR STRUCTURE 211

$$\begin{array}{cccccccc} -7 & -5 & -3 & -1 & +1 & +3 & +5 & +7 \\ +7 & +1 & -3 & -5 & -5 & -3 & -1 & -7 \end{array}$$

and for $n = 9$

$$\begin{array}{ccccccccc} -4 & -3 & -2 & -1 & 0 & +1 & +2 & +3 & +4 \\ +28 & +7 & -8 & -17 & -20 & -17 & -8 & +7 & +28. \end{array}$$

Now for $v = 3$, $n = 9$ division into three blocks followed by randomization and the fitting of (8.31) would have treatment assignments some of which are quite inefficient; moreover no convincing randomization justification would be available. The same would apply *a fortiori* to $v = 4$, $n = 8$ and with rather less force to $v = 2$, $n = 8$. This suggests that, subject to reasonableness in the specific context, it is sensible in such situations to base both design and analysis directly on the model (8.31).

Then a design such that the least squares analysis of (8.31) is associated with a diagonal matrix, i.e. such that the coefficient vectors associated with treatments are orthogonal to the vector of orthogonal polynomials, will generate estimates of minimum variance.

This suggests that we consider the $v \times p$ matrix W^* with elements

$$w^*_{iq} = \Sigma_{T_i} \phi^*_q(s), \qquad (8.32)$$

where the sum is over all units s assigned to T_i and the asterisks refer to orthogonal polynomials normalized to have unit sum of squares over the data points. The objective is to make $W^* = 0$ and where that is not combinatorially possible to make W^* small in some sense. Some optimality requirements that can be used to guide that choice are described in Chapter 7. In some contexts some particular treatment contrasts merit special emphasis. Often, however, there results a design with all the elements w^*_{iq} small and the choice of specific optimality criterion is not critical.

The form of the optimal solution can be seen by examining first the special case of two treatments each replicated r times, $n = 2r$. We take the associated linear model in the slightly modified form

$$E(Y_s) = \mu \pm \delta + \beta_1 \phi^*_1(s) + \ldots + \beta_p \phi^*_p(s), \qquad (8.33)$$

where the treatment effect is 2δ and the normalized form of the orthogonal polynomials is used. The matrix determining the least squares estimates and their precision is

$$\begin{pmatrix} n & 0 & 0 & 0 & \cdots & 0 \\ 0 & n & d_1^* & d_2^* & \cdots & d_p^* \\ 0 & d_1^* & 1 & 0 & \cdots & 0 \\ \cdot & \cdot & \cdot & \cdot & & \cdot \\ 0 & d_p^* & 0 & 0 & \cdots & 1 \end{pmatrix}, \qquad (8.34)$$

where

$$d_q^* = \Sigma_{T_2} \phi_q^*(i) - \Sigma_{T_1} \phi_q^*(i). \qquad (8.35)$$

The inverse of the bordered matrix is easily calculated and from it the variance of $2\hat{\delta}$, the estimated treatment effect, is

$$\frac{2\sigma^2}{r}(1 - \frac{\Sigma d_t^{*2}}{2r})^{-1}. \qquad (8.36)$$

Some of the problems with the approach are illustrated by the case $n = 8$. For $p = 2$ the design

$$T_2 T_1 T_1 T_2 T_1 T_2 T_2 T_1$$

achieves exact orthogonality, $d_1^* = d_2^* = 0$ so that quadratic trend elimination and estimation is achieved without loss of efficiency. On the other hand the design is sensitive to cubic trends and if a cubic trend is inserted into the model by introducing $\phi_3(s)$ the variance of the estimated treatment effect is almost doubled, i.e. the design is only 50% efficient. A design specifically chosen to minimize variance for $p = 3$ has about 95% efficiency.

For general values of the number, v, of treatments similar arguments hold. The technically optimal design depends on the specific criterion and for example on whether some treatment contrasts are of more concern than others, but usually the choice for given p is not critical.

Suppose next that the error structure forms a stationary time series. We deal only with a first-order autoregressive form in which the Gaussian log likelihood is

$$-\log\sigma - (n-1)\log\omega - (y_1 - \mu - \tau_{[1]})^2/(2\sigma^2)$$
$$-\Sigma\{y_s - \mu - \tau_{[s]} - \rho(y_{s-1} - \mu - \tau_{[s-1]})\}^2/(2\omega)^2, \qquad (8.37)$$

where $\tau_{[s]}$ is the treatment effect for the treatment applied to unit s, where ρ is the correlation parameter of the autoregressive process and where σ^2 and ω^2 are respectively marginal and innovation variances of the process so that $\sigma^2 = \omega^2/(1-\rho^2)$. The formally

DESIGNS FOR ONE-DIMENSIONAL ERROR STRUCTURE   213

anomalous role of the first error component arises because it is assumed to have the stationary distribution of the process.

Maximum likelihood can now be applied to estimate the parameters. It is plausible, and can be confirmed by detailed calculation, that at least for $\rho \geq 0$, the most common case, a suitable design strategy is to aim for *neighbourhood balance*. That is, every pair of different treatments should occur next to one another the same number of times. Note, however, that for $\rho < 0$ allocating the same treatment to some pairs of adjacent units enhances precision.

It is instructive and of intrinsic interest to consider the limiting case $\rho = 1$ corresponding to a random walk error structure. In this the error component for the sth experimental unit is

$$\zeta_1 + \ldots + \zeta_s,$$

where the $\{\zeta_s\}$ are uncorrelated random variables of zero mean and variance σ_ζ^2, say. Suppose that T_u is adjacent to T_s λ_{su} times, for $s < u$. We may replace the full set of responses by the first response and the set of differences between adjacent responses. Under the proposed error structure the resulting errors are uncorrelated. Moreover the initial response is the only one to have an expectation depending on the overall mean μ and hence is uninformative about treatment contrasts. Thus the design is equivalent to an incomplete block design with two units per block and with no possibility of recovery of interblock information. When the treatments are regarded symmetrically it is known that a balanced incomplete block design is optimal, i.e. that neighbourhood balance is the appropriate design criterion.

Suppose that this condition is exactly or nearly satisfied, so that there are λ occurrences of each pair of treatments. Then with v treatments there are $n = \lambda v(v-1)/2$ units in the effective incomplete block design and thus $r = \lambda(v-1)/2$ replicates of each treatment. Note, however, that in the notional balanced incomplete design associated with the system each treatment is replicated $2r$ times, because each unit contributes to two differences. The efficiency factor for the balanced incomplete block design is $v/\{2(v-1)\}$ so that the variance of the difference between two treatments is, from Section 4.2.4,

$$\sigma_\zeta^2(v-1)/(vr). \tag{8.38}$$

Suppose now that the system is investigated by a standard randomized block design with its associated randomization analysis.

Within one block the unit terms are of the form

$$\zeta_1, \ \zeta_1 + \zeta_2, \ \ldots, \ \zeta_1 + \ldots + \zeta_v \tag{8.39}$$

and the effective error variance for the randomized block analysis is the expected mean square within this set, namely

$$\sigma_\zeta^2 (v+1)/6. \tag{8.40}$$

Thus the variance of the estimated difference between two treatments is

$$\sigma_\zeta^2 (v+1)/(3r)$$

so that the asymptotic efficiency of the standard randomized block design and analysis is

$$3(v-1)/\{v(v+1)\}. \tag{8.41}$$

Thus at $v = 2$ the efficiency is $1/2$, as is clear from the consideration that half the information contained in a unit is lost in interblock contrasts. The efficiency is the same at $v = 3$ and thereafter decreases slowly with v. The increase in effective variance and decrease in efficiency factor as v increases are partially offset by the decreasing loss from interblock comparisons.

A possible compromise for larger v, should it be desired to use a standard analysis with a clear randomization justification, is to employ a balanced incomplete block design with the number of units per block two or three, i.e. chosen to optimize efficiency if the error structure is indeed of the random walk type. Another possibility in a very large experiment is to divide the experiment into a number of independent sections with a systematic neighbour balance design within each section and to randomize the names of the treatments separately in the different sections.

Finally we need to give sequences that have neighbour balance. See Appendix B for some of the underlying algebra. For two, three and four treatments suitable designs iterate the initial sequences

$$T_1, T_2$$
$$T_1, T_2, T_3; T_2, T_1, T_3;$$
$$T_1, T_2, T_3, T_4; T_2, T_3, T_1, T_4; T_3, T_1, T_4, T_2.$$

There is an anomaly arising from end effects which prevent the balance condition being exactly satisfied unless the designs are regarded as circular, i.e. with the last unit adjacent to the first, but this is almost always not meaningful. In a large design with many repetitions of the above the end effect is negligible.

8.5 Spatial designs

We now turn briefly to situations in which an important feature is the arrangement of experimental units in space. In so far as special models are concerned the discussion largely parallels that of the previous section on designs with a one-dimensional array of units. A further extension, unexplored so far as we are aware, would be to spatial-temporal arrangements of experimental units.

There are two rather different situations. In the first one or more compact areas of ground are available and the issue is to divide them into subareas whose size and shape are determined by the investigator, usually constrained by technological considerations of ease of processing and often also by the need for guard areas to isolate the distinct units. Except for the guard areas the whole of the available area or areas is used for the experiment. The other possibility is that a very large area is available within which relatively small subareas are chosen to form the experimental units.

Illustration. In an agricultural field trial one or more areas of land are divided into plots, the area and size of these being determined in part by ease of sowing and harvesting. By contrast in an ecological experiment a large area of, say, forest is available. Within selected areas different treatments for controlling disease are to be compared. With k treatments for comparison, a version of a randomized block design will require the definition of a number of sets of k areas. The areas within a set should be close together to achieve homogeneity, although sufficiently separated to ensure no leakage of treatment from one area to another and no direct transmission of infection from one area to another.

For example with $k = 3$ the areas might be taken as circles of radius r centred at the corners of an equilateral triangle of side $d, d > 2r$. The orientation of the triangle might not be critical; the centroids of different triangles would be chosen to sample a range of terrains and perhaps to ensure good representation of regions of potentially high infection, in order to study treatment differences in a context where most sensitive estimation of effects is possible.

In a recent experiment on the possible role of badgers in bovine tuberculosis the experimental areas were clusters of three approximately circular areas of radius about 10 km with separation zones.

The treatments were proactive culling of badgers, reactive culling following a breakdown, and no culling. The regions were chosen initially as having high expected breakdown rates on the basis of past data. Before randomization the regions were modified to avoid major rivers, motorways, etc., except as boundaries. The whole investigation consists of ten such triplets, thus forming a randomized block design of ten blocks with three units per block.

When the units are formed from a set of essentially contiguous plots a key traditional design is the Latin square or some generalization thereof. It is assumed that the plots are oriented so that any predominant patterns of variability lie along the rows and columns and not diagonally across the square.

There are many variants of the design when the number of treatments is too large to fit into the simple Latin square form or, possibly, that even one full Latin square would involve too much replication of each treatment. Youden squares provide one extension in which the rows, say, form a balanced incomplete block design and each treatment continues to fall once in each column. We shall describe only one further possibility, the lattice square designs.

In these designs the whole design consists of a set of $q \times q$ squares, where the number of treatments is q^2. The design is such that each treatment occurs once in each square and each pair of treatments occurs together in the same row or column of a square the same number of times. Such designs exist when q is a prime power, p^m, say. The construction is based on the existence of a complete set of $(q-1)$ mutually orthogonal Latin squares.

We first set out the treatments in a $q \times q$ square called a key pattern. Each treatment in effect has attached to it $(q+1)$ aspects, row number, column number and the $(q-1)$ symbols in the Galois field $\mathrm{GF}(p^m)$ in the various mutually orthogonal Latin squares. Notionally we may use the Galois field symbols to label also the rows and columns. We then form squares by taking the labelling characteristics in pairs, choosing each equally often. Thus if q is even, we need to take each twice and if q is odd, so that the number of labelling characteristics is even, then each can be used once.

Table 8.2 shows the design for nine treatments in two 3×3 squares. The first part of the table gives the key pattern and two orthogonal 3×3 Latin squares. Imagine the rows and columns labelled (0, 1, 2). The design itself, before randomization, is formed by taking (rows, columns) and (first alphabet, second alphabet) as

Table 8.2 *3 × 3 lattice squares for 9 treatments: (a) key pattern and orthogonal Latin squares; (b) the design.*

			(a)			
1	2	3		00	11	22
4	5	6		12	20	01
7	8	9		21	02	10

			(b)			
1	2	3		1	6	8
4	5	6		9	2	4
7	8	9		5	7	3

determining via the key pattern the treatments to be assigned to any particular cell. For instance, the entry in row 1 and column 0 of the second square is the element in the key pattern corresponding to $(1, 0)$ in the two orthogonal squares. The way that the design is constructed ensures that each pair of treatments occurs together in either a row or a column just once. The design is resolvable.

For 16 treatments the rows and columns of the key pattern and the three alphabets of the orthogonal Latin squares give five criteria. In this case it needs five 4×4 squares to achieve balance, for example via (row, column); (column, alphabet 1); (alphabet 1, alphabet 2); (alphabet 2, alphabet 3); (alphabet 3, row).

The above designs all have what might be called a traditional justification. That is, for continuous responses approximately normally distributed there is a naturally associated linear model with a justification via randomization theory. For other kinds of response, for example binary responses, it will often be reasonable to start with the corresponding exponential family generalization.

While the Latin square and similar designs retain some validity whatever the pattern of uncontrolled variability they are sensible designs when any systematic effects are essentially along the rows and columns. Occasionally more specific assumptions may be suitable. Thus suppose that we have a spatial coordinate system (η, ζ) corresponding to the rows and columns and that the uncontrolled component of variation associated with the unit centred at (η, ζ) has the generalized additive form

$$a(\eta) + b(\zeta) + \epsilon(\eta, \zeta), \tag{8.42}$$

Table 8.3 *4×4 Latin square. Formal cross-product values of linear by linear components of variation and an optimal treatment assignment.*

$+9, T_1$	$+3, T_2$	$-3, T_3$	$-9, T_4$
$+3, T_4$	$+1, T_3$	$-1, T_2$	$-3, T_1$
$-3, T_3$	$-1, T_4$	$+1, T_1$	$+3, T_2$
$-9, T_2$	$-3, T_1$	$+3, T_4$	$+9, T_3$

where the ϵ's are independent and identically distributed and $a(\eta)$ and $b(\zeta)$ are arbitrary functions. Then clearly under unit-treatment additivity the precision of estimated treatment contrasts is determined by $\text{var}(\epsilon)$.

Now suppose instead that the variation is, except for random error, a polynomial in (η, ζ). We consider a second degree polynomial. Because of the control over row and column effects, a Latin square design balances out all terms except the cross-product term $\eta\zeta$. Balance of the pure linear and quadratic terms does not require the strong balance of the Latin square but for simplicity we restrict ourselves to Latin squares and look for that particular square which is most nearly balanced with respect to the cross-product term.

The procedure is illustrated in Table 8.3. The rows and columns are identified by the standardized linear polynomial with equally spaced levels, for example by -3, -1, 1, 3 for a 4×4 square. With the units of the square are then associated the formal product of the row and column identifiers. By trial and error we find the Latin square most nearly balanced with respect to the cross-product term. This is shown for a 4×4 square in Table 8.3. Especially if the square is to be replicated, the names of the treatments should be randomized within each square but additional randomization would destroy the imposed balance.

The subtotals of the cross-product terms for the four treatments are respectively $4, -4, 4, -4$. These would be zero if exact orthogonality between the cross-product spatial term and treatments could be achieved. In fact the loss of efficiency from the nonorthogonality is negligible, much less than if a Latin square had been randomized. Note that if the four treatments were those in a 2^2 factorial system it would be possible to concentrate the loss of efficiency on the interaction term.

If the design were analysed on the basis of the quadratic model of uncontrolled variation, two degrees of freedom would be removed from each of the between rows and between columns sums of squares and reallocated to the residual.

Often, for example in agricultural field trials, a much more realistic model of spatial variation can be based on a stochastic model of neighbourhood dependence. The simplest models of such type regard the units centred at $(\eta-1,\zeta),(\eta+1,\zeta),(\eta,\zeta-1),(\eta,\zeta+1)$ as the neighbours $N(\eta,\zeta)$ of the unit centred at (η,ζ). If $\xi(\eta,\zeta)$ is the corresponding component of uncontrolled variation one representation of spatially correlated variation has

$$\xi(\eta,\zeta) - \alpha \Sigma_{j \in N(\eta,\zeta)} \xi(j) = \epsilon(\eta,\zeta), \tag{8.43}$$

where the ϵ's are independently and identically distributed.

A different assumption is that the ξ's are generated by a two-dimensional Brownian motion, i. e. that

$$\xi(\eta,\zeta) = \Sigma_{\eta' \leq \eta, \zeta' \leq \zeta} \epsilon^*(\eta',\zeta'), \tag{8.44}$$

where again the ϵ^*'s are independent and identically distributed.

It is in both cases then very appealing to look for designs in which every pair of treatments are neighbours of one another the same number of times.

Sometimes, especially perhaps when a rather inappropriate design has been used, it may, whatever the design, be reasonable to fit a realistic spatial model, for example with a long-tailed distribution of ϵ or ϵ^*. This may partly recover the efficiency probably achievable more simply by more appropriate design. Typically quite extensive calculations, for example by Markov chain Monte Carlo methods, will be needed. Also the conclusions will often be relatively sensitive to the assumptions about the uncontrolled variation.

8.6 Bibliographic notes

There is a very extensive literature on the choice of sample size via considerations of power. A thorough account is given by Desu and Raghavarao (1990). For an early decision-oriented analysis, see Yates (1952).

Optimal stopping in a sequential decision-making formulation is connected with general sequential decision making; for formulations aimed at clinical trials see, for example, Carlin, Kadane and Gelfand (1998) and Wang and Leung (1998).

For a critique of enrichment entry, see Leber and Davis (1998).

Most designs in which the treatment is adapted sequentially trial by trial have little or no element of randomization. For discussion of a design in which randomization is needed and an application in psychophysics, see Rosenberger and Grill (1997).

There is a very extensive literature on sequential stopping, stemming originally from industrial inspection (Wald, 1947) and more recently motivated largely by clinical trials (Armitage, 1975; Whitehead, 1997; Jennison and Turnbull, 2000).

Early work (Neyman, 1923; Hald, 1948) on designs in the presence of polynomial trends presupposed a systematic treatment arrangement with a number of replicates of the same sequence. Cox (1951) discussed the choice of arrangements with various optimum properties and gave formulae for the increase in variance consequent on nonorthogonality. Atkinson and Donev (1996) have reviewed subsequent developments and given extensions. Williams (1952) discussed design in the presence of autocorrelated error structure and Kiefer (1958) proved the optimality of Williams's designs. Similar combinatorial arrangements needed for long sequences on a single subject are mentioned in the notes for Chapter 4.

Methods for using local spatial structure to improve precision in field trials stem from Papadakis (1937); see also Bartlett (1938). Subsequent more recent work, for example Bartlett (1978), makes some explicit use of spatial stochastic models leading to some notion of neighbourhood balance as a design criterion. At the time of writing the extensive literature on the possible advantages of neighbourhood balanced spatial designs over randomized blocks and similar techniques is best approached via the paper of Azaïs, Monod and Bailey (1998) showing how a careful assessment of relative advantage is to be made and the theoretical treatment of randomization theory under a special method of analysis (Monod, Aza+is and Bailey, 1996). Besag and Higdon (1999) describe a very detailed analysis of some spatial designs based on a Markov chain Monte Carlo technique using long-tailed distributions and a specific spatial model. For a general review of the applications to agricultural field trials, see Gilmour, Cullis and Verbyla (1997).

8.7 Further results and exercises

1. Suppose that in comparing two treatments T and C, a variable Y_1 is measured on r_1 units for each treatment with an error variance after allowing for blocking of σ_1^2. It is then decided that a different response variable Y_2 is to be preferred in terms of which the final comparison of treatments is to be made. Therefore a further r_{12} units are assigned to each treatment on which both Y_1 and Y_2 are to be measured followed by a further r_2 units for each treatment on which only Y_2 is measured.

   Under normal theory assumptions in which the regression of Y_2 on Y_1 is the same for both treatments, obtain a likelihood based method for estimating the required treatment effect. What considerations would guide the choice of r_{12} and r_2?

2. The Robbins-Monro procedure is an adaptive treatment assignment rule in effect of the up-and-down type with shrinking step sizes. If we observe a response variable Y with expectation depending in a monotone increasing way on a treatment (dose) variable x via $E(Y; x) = \eta(x)$, where $\eta(\cdot)$ is unknown, the objective is assumed to be the estimation for given p of $x^{(p)}$, where

   $$\eta(x^{(p)}) = p.$$

   Thus for a binary response and $p = 1/2$, estimation is required of the 50% point of the response, the ED50. The procedure is to define treatment levels recursively depending on whether the current response is above or below the target via

   $$x_{t+1} = x_t - a_t(x_t - p),$$

   where the preassigned sequence a_t defines the procedure. If the procedure is stopped after n steps the estimate of $x^{(p)}$ is either x_n or x_{n+1}.

   Give informal motivation for the formal conditions for convergence of the procedure that Σa_t is divergent and Σa_t^2 convergent leading to the common choice $a_n = a/n$, for some constant a. Note, however, that the limiting behaviour as t increases is relevant only as a guide to behaviour in that the procedure will always have to be combined with a stopping rule.

   Assume that the procedure has reached a locally linear region near $x^{(p)}$ in which the response function has slope $\eta'(p)$ and the variance of Y is constant, σ^2, say. Show by rewriting the defining equation in terms of the response Y_t at x_t and assuming appropriate dependence on sample size that the asymptotic variance is

$$a^2\sigma^2/(2a\eta'^{(p)} - 1).$$

How might this be used to choose a?

The formal properties were given in generality by Robbins and Monro (1951); for a more informal discussion see Wetherill and Glazebrook (1986, Chapters 9 and 10).

3. If point events occur in a Poisson process of rate ρ, the number N_t of points in an interval of length t_0 has a Poisson distribution with mean and variance equal to ρt_0. Show that for large t_0, $\log N_{t_0} - \log t_0$ is asymptotically normal with mean $\log \rho$ and variance $1/(\rho t_0)$ estimated by $1/N_{t_0}$.

Show that if, on the other hand, sampling proceeds until a preassigned number n_0 of points have been observed then the corresponding time period T_0 is such that $2\rho T_0$ has a chi-squared distribution with $2n_0$ degrees of freedom and that the asymptotic variance of the estimate of ρ is again the reciprocal of the number of points counted.

Suppose now that two treatments are to be compared with corresponding Poisson rates ρ_1, ρ_2. Show that the variance of the estimate of $\log \rho_2 - \log \rho_1$ is approximately $1/N_2 + 1/N_1$ which if the numbers are not very different is approximately $4/N.$; here N_1 and N_2 are the numbers of points counted in the two groups and $N. = N_1 + N_2$. Hence show that to achieve a certain preassigned fractional standard error, d, in estimating the ratio sampling should proceed until about $4/d^2$ points have been counted in total, distributing the sampling between the two groups to achieve about the same number of points from each.

What would be the corresponding conclusion if both processes had to be corrected for a background noise process of known rate ρ_0? What would be the consequences of overdispersion in the Poisson processes?

4. In a randomized trial to compare two treatments, T and C, with equal replication over $2r$ experimental units, suppose that treatments are allocated to the units in sequence and that at each stage the outcomes of the previous allocations are known and moreover that the strategy of allocation is known. Some aspects

of the avoidance of selection bias can be represented via a two-person zero-sum game in which player I chooses the treatment to be assigned to each individual and player II "guesses" the outcome of the choice. Player II receives from or pays out to player I one unit depending on whether the guess is correct or false. Blackwell and Hodges (1957) show that the design in which the treatments are allocated independently with equal probabilities until one treatment has been allocated r times is the optimal strategy for player I with the obvious associated rule for player II and that the expected number of correct guesses by player II exceeds r by

$$r\binom{2r}{r}/2^{2r} \sim \sqrt{(r/\pi)},$$

whereas if all treatment allocations are equally likely and player II acts appropriately the corresponding excess is

$$2^{2r-1}/\binom{2r}{r} \sim \sqrt{(\pi r/4)}.$$

Note that the number of excess correct guesses could be reduced to zero by independently randomizing each unit but this would carry a penalty in terms of possibly serious imbalance in the two treatment arms.

5. The following adaptive randomization scheme has been used in clinical trials to compare two treatments T and C when a binary response, success or failure, can be observed on each experimental unit before the next experimental unit is to be randomized. The initial randomization is represented by an urn containing two balls marked T and C respectively. Each time a success is observed, a ball marked by the successful treatment is added to the urn.

In a trial on newborn infants with respiratory failure, the new treatment T was highly invasive: extracorporeal membrane oxygenation (ECMO), and C was conventional medical management. The first patient was randomized to T and survived, the second was randomized to C and died, and the next ten patients were randomized to T, all surviving, at which time the trial was terminated. (Wei, 1988; Bartlett et al., 1985).

Compare the efficiency of the adaptive urn scheme to that of balanced allocation to T and C, for an experiment with a total sample of size 12, first under the assumption that p_1, the probability of success under T is 0.80, and p_2, the probability of success under C is 0.20, and then under the assumption that $p_1 = p_2$. See Begg (1990) for a discussion of inference under the adaptive randomization scheme.

The results of this study were considered sufficiently inconclusive that another trial was conducted in Boston in 1986, using a sequential allocation scheme in which patients were randomized equally to T and C in blocks of size four, until four deaths occurred on either T or C. The rationale for this design and the choice of stopping rule is given in Ware (1989); analysis of the resulting data (9 units randomized to T with no failures, 10 units randomized to C with 4 failures) indicates substantial but not overwhelming evidence in favour of ECMO. The discussion of Ware (1989) highlights several interesting ethical and statistical issues.

Subsequent studies have not completely clarified the issue, although the UK Collaborative ECMO Trial (1996) estimated the risk of death for ECMO relative to conventional therapy to be 0.55.

APPENDIX A
Statistical analysis

A.1 Introduction

Design and statistical analysis are inextricably linked but in the main part of the book we have aimed primarily to discuss design with relatively minimal discussion of analysis. Use of results connected with the linear model and analysis of variance is, however, unavoidable and we have assumed some familiarity with these. In this Appendix we describe the essential results required in a compact, but so far as feasible, self-contained way.

To the extent that we concern ourselves with analysis, we represent the response recorded on a particular unit as the value of a random variable and the objective to be inference about aspects of the underlying probability distributions, in particular parameters describing differences between treatments. Such models are an essential part of the more formal part of statistical analysis, i.e. that part that goes beyond graphical and tabular display, important though these latter are.

One of the themes of the earlier chapters of this book is an interplay between two different kinds of probabilistic model. One is the usual one in discussions of statistical inference where such models are idealized representations of physical random variability as it arises when repeat observations are made under nominally similar conditions. The second model is one in which the randomness enters only via the randomization procedure used by the investigator in allocating treatments to experimental units. This leads to the notion that a standard set of assumptions plus consideration of the design used implies a particular form of default analysis without special assumptions about the physical form of the random variability encountered. These considerations are intended to remove some of the arbitrariness that may seem to be involved in constructing models and analyses for special designs.

A.2 Linear model

A.2.1 Formulation and assumptions

We write the linear model in the equivalent forms

$$E(Y) = X\theta, \quad Y = X\theta + \epsilon, \tag{A.1}$$

where by definition $E(\epsilon) = 0$. Here Y is a $n \times 1$ vector of random variables representing responses to be observed, one per experimental unit, θ is a $q \times 1$ vector of unknown parameters representing variously treatment contrasts, including main effects and interactions, block and similar effects, effects of baseline variables, etc. and X is a $n \times q$ matrix of constants determined by the design and other structure of the system. Typically some components of θ are parameters of interest and others are nuisance parameters.

It is frequently helpful to write $q = p+1$ and to take the first column of X to consist of ones, concentrating then on the estimation of the last p components of θ. Initially we suppose that X is of full rank $q < n$. That is there are fewer parameters than observations, so that the model is not saturated with parameters, and moreover there is not a redundant parameterization.

For the primary discussion we alternate between two possible assumptions about the error vector ϵ: it is always clear from context which is being used.

Second moment assumption. The components of ϵ are uncorrelated and of constant variance σ^2, i.e.

$$E(\epsilon\epsilon^T) = \sigma^2 I, \tag{A.2}$$

where I is the $n \times n$ identity matrix.

Normal theory assumption. The components of ϵ are independently normally distributed with zero mean and variance σ^2.

Unless explicitly stated otherwise we regard σ^2 as unknown. The normal theory assumption implies the second moment assumption. The reasonableness of the assumptions needs consideration in each applied context.

A.2.2 Key results

The strongest theoretical motivation of the following definitions is provided under the normal theory assumption by examining the likelihood function, checking that it is the likelihood for an exponential family with $q + 1$ parameters and a $q + 1$ dimensional

LINEAR MODEL

canonical statistic and that hence analysis under the model is to be based on the statistics now to be introduced. We discuss optimality further in Section A2.4 but for the moment simply consider the following statistics.

We define the least squares estimate of θ by the equation

$$X^T X \hat{\theta} = X^T Y, \tag{A.3}$$

the residual vector to be

$$Y_{\text{res}} = Y - X\hat{\theta} \tag{A.4}$$

and the residual sum of squares and mean square as

$$\text{SS}_{\text{res}} = Y_{\text{res}}^T Y_{\text{res}}, \quad \text{MS}_{\text{res}} = \text{SS}_{\text{res}}/(n-q). \tag{A.5}$$

Occasionally we use the extended notation $Y_{\text{res}.X}$, or even $Y_{.X}$, to show the vector and model involved in the definition of the residual.

Because X has full rank so too does $X^T X$ enabling us to write

$$\hat{\theta} = (X^T X)^{-1} X^T Y. \tag{A.6}$$

Under the second moment assumption $\hat{\theta}$ is an unbiased estimate of θ with covariance matrix $(X^T X)^{-1} \sigma^2$ and MS_{res} is an unbiased estimate of σ^2. Thus the covariance matrix of $\hat{\theta}$ can be estimated and approximate confidence limits found for any parametric function of θ. One strong justification of the least squares estimates is that they are functions of the sufficient statistics under the normal theory assumption. Another is that among unbiased estimators linear in Y, $\hat{\theta}$ has the "smallest" covariance matrix, i.e. for any matrix C for which $E(CY) = \theta$, $\text{cov}(\hat{\theta}) - \text{cov}(CY)$ is positive semi-definite. Stronger results are available under the normal theory assumption; for example $\hat{\theta}$ has smallest covariance among all unbiased estimators of θ.

Under the second moment assumption on substituting $Y = X\theta + \epsilon$ into (A.6) we have

$$\begin{aligned} \hat{\theta} &= (X^T X)^{-1} X^T X \theta + (X^T X)^{-1} X^T \epsilon \\ &= \theta + (X^T X)^{-1} X^T \epsilon. \end{aligned} \tag{A.7}$$

The unbiasedness follows immediately and the covariance matrix of $\hat{\theta}$ is

$$\begin{aligned} E\{(\hat{\theta} - \theta)(\hat{\theta} - \theta)^T\} &= (X^T X)^{-1} X^T E(\epsilon \epsilon^T) X (X^T X)^{-1} \\ &= (X^T X)^{-1} \sigma^2. \end{aligned} \tag{A.8}$$

Further
$$Y_{\text{res}} = \{I - X(X^T X)^{-1} X^T\}\epsilon. \tag{A.9}$$

Now the residual sum of squares is $Y_{\text{res}}^T Y_{\text{res}}$, so that the expected value of the residual sum of squares is
$$\sigma^2 \text{tr}(\{I - X(X^T X)^{-1} X^T\}^T \{I - X(X^T X)^{-1} X^T\}).$$

Direct multiplication shows that
$$\{I - X(X^T X)^{-1} X^T\}^T \{I - X(X^T X)^{-1} X^T\}$$
$$= \{I - X(X^T X)^{-1} X^T\}$$

and its trace is
$$\text{tr}\{(I_n - X^T X(X^T X)^{-1})\} = \text{tr}(I_n) - \text{tr}(I_q) = n - q, \tag{A.10}$$

where temporarily we show explicitly the dimensions of the identity matrices involved.

A.2.3 Some properties

There is a large literature associated with the results just sketched and their generalizations. Here we give only a few points.

First under the normal theory assumption the log likelihood is, except for a constant
$$-n \log \sigma - (Y - X\theta)^T (Y - X\theta)/(2\sigma^2). \tag{A.11}$$

The identity
$$(Y - X\theta)^T (Y - X\theta)$$
$$= \{(Y - X\hat{\theta}) + X(\hat{\theta} - \theta)\}^T \{(Y - X\hat{\theta}) + X(\hat{\theta} - \theta)\}$$
$$= \text{SS}_{\text{res}} + (\hat{\theta} - \theta)^T (X^T X)(\hat{\theta} - \theta), \tag{A.12}$$

the cross-product term vanishing, justifies the statement about sufficiency at the beginning of Section A2.2.

Next we define the vector of fitted values
$$\hat{Y} = X\hat{\theta}, \tag{A.13}$$

the values that would have arisen had the data exactly fitted the model with the estimated parameter value. Then we have the analysis of variance, or more literally the analysis of sum of squares,
$$Y^T Y = (Y - X\hat{\theta} + X\hat{\theta})^T (Y - X\hat{\theta} + X\hat{\theta})$$

$$= \text{SS}_{\text{res}} + \hat{Y}^T \hat{Y}. \tag{A.14}$$

We call the second term the sum of squares for fitting X and sometimes denote it by SS_X. It follows on direct substitution that

$$E(\text{SS}_X) = (X\theta)^T(X\theta) + q\sigma^2. \tag{A.15}$$

A property that is often useful in analysing simple designs is that because $X^T X$ is of full rank, a component $\hat{\theta}_s$ of the least squares estimate is the unique linear combination of $X^T Y$, the right-hand side of the least squares equations, that is unbiased for θ_s. For such a linear combination $l_s^T X^T Y$ to be unbiased we need

$$E(l_s^T X^T Y) = l_s^T X^T X \theta = e_s^T \theta, \tag{A.16}$$

where e_s is a vector with one in row s and zero elsewhere. This implies that l_s is the sth column of $(X^T X)^{-1}$.

Finally, and most importantly, consider confidence limits for a component parameter. Write $C = X^T X$ and denote the elements of C^{-1} by c^{rs}. Then

$$\text{var}(\hat{\theta}_s) = c^{ss} \sigma^2$$

is estimated by

$$c^{ss} \text{MS}_{\text{res}}$$

suggesting the use of the pivot

$$(\hat{\theta}_s - \theta)/\sqrt{(c^{ss} \text{MS}_{\text{res}})} \tag{A.17}$$

to calculate confidence limits for and test hypotheses about θ_s.

Under the normal theory assumption the pivot has a Student t distribution with $n - q$ degrees of freedom. Under the second moment assumption it will have for large n asymptotically a standard normal distribution under the extra assumptions that

1. $n - q$ also is large which can be shown to imply that MS_{res} converges in probability to σ^2
2. the matrix X and error structure are such that the central limit theorem applies to $\hat{\theta}_s$.

Over the second point note that if we assumed the errors independent, and not merely uncorrelated, it is a question of verifying say the Lindeberg conditions. A simple sufficient but not necessary condition for asymptotic normality of the least squares estimates is then that in a notional series of problems in which the number of parameters is fixed and the number of observations tends to infinity the squared norms of all columns of $(X^T X)^{-1} X^T$ tend to zero.

A.2.4 Geometry of least squares

For the study of special problems that are not entirely balanced we need implicitly or explicitly either algebraic manipulation and simplification of the above matrix equations or, perhaps more commonly, their numerical evaluation and solution. For some aspects of the general theory, however, it is helpful to adopt a more abstract approach and this we now sketch.

We shall regard the vector Y and the columns of X as elements of a linear vector space $\mathcal{V}$. That is, we can add vectors and multiply them by scalars and there is a zero vector in $\mathcal{V}$.

We equip the space with a norm and a scalar product and for most purposes use the Euclidean norm, i.e. we define for a vector $Z \in \mathcal{V}$, specified momentarily in coordinate form,

$$\|Z\|^2 = Z^T Z = \Sigma Z_i^2, \tag{A.18}$$

and the scalar product by

$$(Z_1, Z_2) = Z_1^T Z_2 = (Z_2, Z_1). \tag{A.19}$$

Two vectors are orthogonal if their scalar product is zero.

Given a set of vectors the collection of all linear combinations of them defines a subspace, $\mathcal{S}$, say. Its dimension $d_\mathcal{S}$ is the maximal number of linearly independent components in $\mathcal{S}$, i.e. the maximal number such that no linear combination is identically zero. In particular the q columns of the $n \times q$ matrix X define a subspace $\mathcal{C}_\mathcal{X}$ called the column space of X. If and only if X is of full rank $q = d_{\mathcal{C}_\mathcal{X}}$.

In the following discussion we abandon the requirement that X is of full rank.

Given a subspace $\mathcal{S}$ of dimension $d_\mathcal{S}$

1. the set of all vectors orthogonal to all vectors in $\mathcal{S}$ forms another vector space $\mathcal{S}^\perp$ called the orthogonal complement of $\mathcal{S}$
2. $\mathcal{S}^\perp$ has dimension $n - d_\mathcal{S}$
3. an arbitrary vector Z in $\mathcal{V}$ is uniquely resolved into two components, its projection in $\mathcal{S}$ and its projection in the orthogonal complement

$$Z = Z_\mathcal{S} + Z_{\mathcal{S}^\perp} \tag{A.20}$$

4. the components are by construction orthogonal and

$$\|Z\|^2 = \|Z_\mathcal{S}\|^2 + \|Z_{\mathcal{S}^\perp}\|^2. \tag{A.21}$$

LINEAR MODEL   231

We now regard the linear model as specifying that $E(Y)$ lies in the column space of X, $\mathcal{C}_\mathcal{X}$. Resolve Y into a component in $\mathcal{C}_\mathcal{X}$ and a component in its orthogonal complement. The first component, in matrix notation $X\tilde{\theta}$, say, is such that the second component $Y - X\tilde{\theta}$ is orthogonal to every column of X, i.e.

$$X^T(Y - X\tilde{\theta}) = 0 \qquad (A.22)$$

and these are the least squares equations (A.3) so that $\tilde{\theta} = \hat{\theta}$. Further, the components are the vectors of fitted values and residuals and the analysis of variance in (A.14) is the Pythagorean identity for their squared norms.

From this representation we have the following results.

In a redundant parameterization, the vector of fitted values and the residual vector are uniquely defined by the column space of X even though some at least of the estimates of individual components of θ are not uniquely defined.

The estimate of a component of θ based on a linear combination of the components of Y is a scalar product (l, Y) of an estimating vector l with Y. We can resolve l into components $l_{\mathcal{C}_\mathcal{X}}, l_{\mathcal{C}_\mathcal{X}}^\perp$ in and orthogonal to $\mathcal{C}_\mathcal{X}$. Every scalar product $(l^\perp, Y)$, in a slightly condensed notation, has zero mean and, because of orthogonality, $\text{var}\{(l, Y)\}$ is the sum of the variances of the components. It follows that for a given expectation the variance is minimized by taking only estimating vectors in $\mathcal{C}_\mathcal{X}$, i.e. by linear combinations of $X^T Y$, justifying under the second moment assumption the use of least squares estimates. This property may be called the *linear sufficiency* of $X^T Y$.

We now sketch the distribution theory underlying confidence limits and tests under the normal theory assumption. It is helpful, although not essential, to set up in $\mathcal{C}_\mathcal{X}$ and its orthogonal complement a set of orthogonal unit vectors as a basis for each space in terms of which any vector may be expressed. By an orthogonal transformation the scalar product of Y with any of these vectors is normally distributed with variance σ^2 and scalar products with different basis vectors are independent. It follows that

1. the residual sum of squares SS_{res} has the distribution of σ^2 times a chi-squared variable with degrees of freedom $n - d_{\mathcal{C}_\mathcal{X}}$

2. the residual sum of squares is independent of the least squares estimates, and therefore of any function of them

3. the least squares estimates, when uniquely defined, are normally distributed

4. the sum of squares of fitted values SS_X has the form of σ^2 times a noncentral chi-squared variable with $d_{\mathcal{C}_X}$ degrees of freedom, reducing to central chi-squared if and only if $\theta = 0$, i.e. the true parameter value is at the origin of the vector space.

These cover the distribution theory for standard so-called exact tests and confidence intervals.

In the next subsection we give a further development using the coordinate-free vector space approach.

A.2.5 Stage by stage fitting

In virtually all the applications we consider in this book the parameters and therefore the columns of the matrix X are divided into sections corresponding to parameters of different types; in particular the first parameter is usually associated with a column of one's, i.e. is a constant for all observations.

Suppose then that

$$E(Y) = X_0 \theta_{0.1} + X_1 \theta_{1.0}; \qquad (A.23)$$

no essentially new ideas are involved with more than two sections. We suppose that the column spaces of X_1 and of X_0 do not coincide and that in general each new set of parameters genuinely constrains the previous model.

It is then sometimes helpful to argue as follows.

Set $\theta_{1.0} = 0$. Estimate θ_0, the coefficient of X_0 in the model ignoring X_1, by least squares. Note that the notation specifies what parameters are included in the model. We call the resulting estimate $\hat{\theta}_0$ the least squares estimate of θ_0 ignoring X_1 and the associated sum of squares of fitted values, SS_{X_0}, the sum of squares for X_0 ignoring X_1.

Now project the whole problem into the orthogonal complement of $\mathcal{C}_{X_0}$, the column space of X_0. That is, we replace Y by what we now denote by $Y_{.0}$, the residual vector with respect to X_0 and we replace X_1 by $X_{1.0}$ and the linear model formally by

$$E(Y_{.0}) = X_{1.0} \theta_{1.0}, \qquad (A.24)$$

a linear model in a space of dimension $n - d_0$, where d_0 is the dimension of $\mathcal{C}_{X_0}$.

LINEAR MODEL   233

We again obtain a least squares estimate $\hat{\theta}_{1.0}$ by orthogonal projection. We obtain also a residual vector $Y_{.01}$ and a sum of squares of fitted values which we call the sum of squares adjusted for X_0, $SS_{X_{1.0}}$.

We continue this process for as many terms as there are in the original model.

For example if there were three sections of the matrix X; X_0, X_1, X_2, the successive sums of squares generated would be for fitting first X_0 ignoring (X_1, X_2), then X_1 ignoring X_2 adjusting for X_0 and finally X_2 adjusting for (X_0, X_1), leaving a sum of squares residual to the whole model. These four sums of squares, being squared norms in orthogonal subspaces, are under the normal theory assumption independently distributed in chi-squared distributions, central for the residual sum of squares and in general noncentral for the others. Although it is an aspect we have not emphasized in the book, if a test is required of consistency with $\theta_{2.01} = 0$ in the presence of arbitrary θ_0, θ_1 this can be achieved via the ratio of the mean square for X_2 adjusting for (X_0, X_1) to the residual mean square. The null distribution, corresponding to the appropriately scaled ratio of independent chi-squared variables, is the standard variance ratio or F distribution with degrees of freedom the dimensions of the corresponding spaces.

The simplest special case of this procedure is the fitting of the linear regression

$$E(Y_i) = \theta_{0.1} + \theta_{1.0} z_i, \qquad (A.25)$$

so that X_0, X_1 are both $n \times 1$. We estimate θ_0 ignoring X_1 by the sample mean $\bar{Y}_.$ and projection orthogonal to the unit vector X_0 leads to the formal model

$$E(Y_i - \bar{Y}_.) = \theta_{1.0}(z_i - \bar{z}_.) \qquad (A.26)$$

from which familiar formulae for a least squares slope, and associated sum of squares, follow immediately.

In balanced situations, such as a randomized block design, in which the three sections correspond to terms representing general mean, block and treatment effects, the spaces $X_{1.0}, X_{2.0}$ are orthogonal and $X_{2.0}$ is the same as $X_{2.01}$. Then and only then the distinction between, say, a sum of squares for blocks ignoring treatments and a sum of squares for blocks adjusting for treatments can be ignored.

Because of the insight provided by fitting parameters in stages and of its connection with the procedure known in the older literature as analysis of covariance, we now sketch an algebraic discussion, taking for simplicity a model in two stages and assuming formulations of full rank.

That is we start again from

$$E(Y) = X_0\theta_{0.1} + X_1\theta_{1.0}, \tag{A.27}$$

where X_j is $n \times d_j$ and $\theta_{j.k}$ is $d_j \times 1$. The matrix

$$R_0 = I - X_0(X_0^T X_0)^{-1} X_0^T \tag{A.28}$$

is symmetric and idempotent, i.e. $R_0^T R_0 = R_0^2 = R_0$ and for any vector u of length n, $R_0 u$ is the vector of residuals after regressing u on X_0. The associated residual sum of squares is $(R_0 u)^T (R_0 u) = u^T R_0 u$ and the associated residual sum of products for two vectors u and v is $u^T R_0 v$. Further for any vector u, we have $X_0^T R_0 u = 0$. In the notation of (A.24) $R_0 Y = Y_{.0}$ and $R_0 X_1 = X_{1.0}$.

We can rewrite model (A.27) in the form

$$E(Y) = X_0\theta_0 + R_0 X_1 \theta_{1.0} \tag{A.29}$$

where

$$\theta_0 = \theta_{0.1} + (X_0^T X_0)^{-1} X_0^T X_1 \theta_{1.0}. \tag{A.30}$$

The least squares equations for the model (A.29) have the form

$$\begin{pmatrix} X_0^T X_0 & 0 \\ 0 & X_1^T R_0 X_1 \end{pmatrix} \begin{pmatrix} \hat{\theta}_0 \\ \hat{\theta}_{1.0} \end{pmatrix} = \begin{pmatrix} X_0^T Y \\ X_1^T R_0 Y \end{pmatrix} \tag{A.31}$$

from which the following results can be deduced.

First the parameters θ_0 and $\theta_{1.0}$ are orthogonal with respect to the expected Fisher information in the model, and the associated vectors X_0 and $X_{1.0}$ are orthogonal in the usual sense.

Secondly $\theta_{1.0}$ is estimated from a set of least squares equations formed from the matrix of sums and squares of products of the columns of X_1 residual to their regression on X_0 and the covariance matrix of $\hat{\theta}_{1.0}$ is similarly determined.

Thirdly the least squares estimate $\hat{\theta}_{1.0}$ is obtained by adding to $\hat{\theta}_0$ a function uncorrelated with it.

Continuing with the analysis of (A.29) we see that the residual sum of squares from the full model is obtained by reducing the residual sum of squares from the model ignoring θ_1, i.e. $Y^T R_0 Y$, by the sum of squares of fitted values in the regression of $R_0 Y$ on

LINEAR MODEL

$R_0 X_1$:

$$(Y - X_0\hat{\theta}_0 - R_0 X_1 \hat{\theta}_{1.0})^T (Y - X_0\hat{\theta}_0 - R_0 X_1 \hat{\theta}_{1.0})$$
$$= Y^T R_0 Y - \hat{\theta}_{1.0}^T X_1^T R_0 X_1 \hat{\theta}_{1.0}. \qquad (A.32)$$

Under the normal theory assumption, an F test of the hypothesis that $\theta_{1.0} = 0$ is obtained via the calculation of residual sums of squares from the full model and from one in which the term in $\theta_{1.0}$ is omitted.

It is sometimes convenient to display the results from stage by stage fitting in an analysis of variance table; an example with just two stages is outlined in Table A.1. In the models used in this book, the first section of the design matrix is always a column of 1's corresponding to fitting an overall mean. This is so rarely of interest it is usually omitted from the analysis of variance table, so that the total sum of squares is $(Y - \bar{Y}1)^T(Y - \bar{Y}1)$ instead of $Y^T Y$.

A.2.6 Redundant parameters

In the main text we often use formulations with redundant parameters, such as for the completely randomized and randomized block

Table A.1 *Analysis of variance table emphasizing the fitting of model (A.27) in two stages.*

Source	D.f.	Sums of squares	Expected mean square
Regr. on X_0	rank X_0 = q	$\hat{\theta}_0^T X_0^T X_0 \hat{\theta}_0$	$\sigma^2 + \theta_0^T X_0^T X_0 \theta_0 / q$
Regr. on X_1, adj. for X_0	$p - q$	$\hat{\theta}_{1.0}^T X_{1.0}^T X_{1.0} \hat{\theta}_{1.0}$	$\sigma^2 + \theta_{1.0}^T X_{1.0}^T X_{1.0} \theta_{1.0} / (p - q)$
Residual	$n - p - q$	$(Y - \hat{Y})^T (Y - \hat{Y})$	σ^2
Total	n	$Y^T Y$	

designs

$$E(Y_{js}) = \mu + \tau_j, \qquad (A.33)$$
$$E(Y_{js}) = \mu + \tau_j + \beta_s. \qquad (A.34)$$

With v treatments and b blocks these models have respectively $v+1$ and $v+b+1$ parameters although the corresponding X matrices in the standard formulation have ranks v and $v+b-1$.

The least squares equations are algebraically consistent but do not have a unique solution. The geometry of least squares shows, however, that the vector of fitted values and the sum of squares for residual and for fitting X are unique. In fact any parameter of the form $a^T X\theta$ has a unique least squares estimate.

From a theoretical viewpoint the most satisfactory approach is via the use of generalized inverses and the avoidance of explicit specification of constraints. The issue can, however, always be by-passed by redefining the model so that the column space of X remains the same but the number of parameters is reduced to eliminate the redundancy, i.e. to restore X to full rank. When the parameters involved are parameters of interest this is indeed desirable in that the primary objective of the analysis is the estimation of, for example, treatment contrasts, so that formulation in terms of them has appeal despite a commonly occurring loss of symmetry.

While, as noted above, many aspects of the estimation problem do not depend on the particular reparameterization chosen some care is needed. First and most importantly the constraint must not introduce an additional rank reduction in X; thus in (A.33) the constraint

$$\mu + \Sigma \tau_s = a \qquad (A.35)$$

for a given constant a would not serve for resolution of parameter redundancy.

In general suppose that the $n \times q$ matrix X is of rank $q - r$ and impose constraints $A\theta = 0$, where A is $r \times q$ chosen so that the equations

$$X\theta = k, \quad A\theta = 0$$

uniquely determine θ for all constant vectors k. This requires that (X^T, A^T) is of full rank. The least squares projection equations

LINEAR MODEL

supplemented by the constraints give

$$\begin{pmatrix} X^T X & A^T \\ A & 0 \end{pmatrix} \begin{pmatrix} \hat{\theta} \\ \lambda \end{pmatrix} = \begin{pmatrix} X^T Y \\ 0 \end{pmatrix}, \qquad (A.36)$$

where λ is a vector of Lagrange multipliers used to introduce the constraint.

The matrix on the left-hand side is singular but is converted into a nonsingular matrix by changing $X^T X$ to $X^T X + A^T A$ which does not change the solution. Equations for the constrained estimates and their covariance matrix follow.

In many of the applications discussed in the book some components of the parameter vector correspond to qualitative levels of a factor, i.e. each level has a distinct component. As discussed above constraints are needed if a model of full rank is to be achieved. Denote the unconstrained set of relevant components by $\{\psi_1, \ldots, \psi_k\}$; typical constraints are $\psi_1 = 0$ and $\Sigma \psi_j = 0$, although others are possible. The objective is typically the estimation with standard errors of a set $\mathcal{C}$ of contrasts $\Sigma c_j \psi_j$ with $\Sigma c_j = 0$. If $\mathcal{C}$ is the set of simple contrasts with baseline, i.e. the estimation of $\psi_j - \psi_1$ for $j = 2, \ldots, k$, the resulting estimates of the constrained parameters and their standard errors are all that is needed. In general, however, the full covariance matrix of the constrained estimates will be required; for example if the constrained parameters are denoted by $\phi_2, \ldots, \phi_k$, estimation of $\psi_3 - \psi_2$ via $\hat{\phi}_3 - \hat{\phi}_2$ is direct but the standard error requires $\text{cov}(\hat{\phi}_3, \hat{\phi}_2)$ as well as the separate variances.

In the presentation of conclusions and especially for moderately large k the recording of a full covariance matrix may be inconvenient. This may often be avoided by the following device. We attach pseudo-variances $\omega_1, \ldots, \omega_k$ to estimates of so-called *floating parameters* $a + \psi_1, \ldots, a + \psi_k$, where a is an unspecified constant, in such a way that exactly or approximately

$$\text{var}(\Sigma c_j \hat{\psi}_j) = \text{var}\{\Sigma c_j (a + \hat{\psi}_j)\} = \Sigma \omega_j c_j^2 \qquad (A.37)$$

for all contrasts in the set $\mathcal{C}$. We then treat the floating parameters as if independently estimated. In simple cases this reduces to specifying marginal means rather than contrasts.

A.2.7 Residual analysis and model adequacy

There is an extensive literature on the direct use of the residual vector, $Y_{\text{res}.X}$, for assessing model adequacy and detecting possible data anomalies. The simplest versions of these methods hinge on the notion that if the model is reasonably adequate the residual vector is approximately a set of independent and identically distributed random variables, so that structure detected in them is evidence of model inadequacy. This is broadly reasonable when the number of parameters fitted is small compared with the number of independent observations. In fact the covariance matrix of the residual vector is

$$\sigma^2 \{I - X(X^T X)^{-1} X^T\} \tag{A.38}$$

from which more refined calculations can be made.

For many of the designs considered in this book, however, q is not small compared with n and naive use of the residuals will be potentially misleading.

Possible departures from the models can be classified roughly as systematic changes in structure, on the whole best detected and analysed by fitting extended models, and data anomalies, such as defective observations, best studied via direct inspection of the data, where these are not too extensive, and by appropriate functions of residuals. See the Bibliographic notes and Further results and exercises.

A.3 Analysis of variance

A.3.1 Preliminaries

In the previous section analysis of variance was introduced in the context of the linear model as a schematic way first of calculating the residual sum of squares as a basis for estimating residual variance and then as a device for testing a null hypothesis constraining the parameter vector of the linear model to a subspace. There are, however, other ways of thinking about analysis of variance.

The first corresponds in a sense to the most literal meaning of analysis. Suppose that an observed random variable is in fact the sum of two unobserved (latent) variables, so that in the simplest case in which systematic effects are absent we can write

$$Y = \mu + \xi + \eta, \tag{A.39}$$

ANALYSIS OF VARIANCE

where μ is the unknown mean and ξ and η are uncorrelated random variables of zero mean and variances respectively $\sigma_\xi^2, \sigma_\eta^2$; these are called *components of variance*. Then

$$\text{var}(Y) = \sigma_\xi^2 + \sigma_\eta^2, \tag{A.40}$$

with obvious generalization to more than two components of variability.

If now we observe a number of different random variables with this general structure it may be possible to estimate the components of variance $\sigma_\xi^2, \sigma_\eta^2$ separately and in this sense to have achieved a splitting up, i.e. analysis, of variance. The two simplest cases are where we have repeat observations on a number of groups of observations Y_{aj}, say, with

$$Y_{aj} = \mu + \xi_a + \eta_{aj} \tag{A.41}$$

for $a = 1, \ldots, A; j = 1, \ldots, J$, where it is convenient to depart briefly from our general convention of restricting upper case letters to random variables.

The formulation presupposes that the variation between groups, as well as that within groups, is regarded as best represented by random variables; it would thus not be appropriate if the groups represented different treatments.

A second possibility is that the observations are cross-classified by rows and columns in a rectangular array and where the observations Y_{ab} can be written

$$Y_{ab} = \mu + \xi_a + \eta_b + \epsilon_{ab}, \tag{A.42}$$

where the component random variables are now interpreted as random row and column effects and as error, respectively.

It can be shown that in these and similar situations with appropriate definitions the components of variance can be estimated via a suitable analysis of variance table. Moreover we can then, via the complementary technique of *synthesis of variance*, estimate the properties of systems in which either the variance components have been modified in some way, or in which the structure of the data is different, the variance components having remained the same.

For example under model (A.41) the variance of the overall mean of the data, considered as an estimate of μ is easily seen to be

$$\sigma_\xi^2/A + \sigma_\eta^2/(AJ).$$

We can thus estimate the effect on the precision of the estimate of

μ of, say, increasing or decreasing J or of improvements in measurement technique leading to a reduction in σ_η^2.

A.3.2 Cross-classification and nesting

We now give some general remarks on the formal role of analysis of variance to describe relatively complicated structures of data. For this we consider data classified in general in a multi-dimensional array. Later it will be crucial to distinguish classification based on a treatment applied to the units from that arising from the intrinsic structure of the units but for the moment we do not make that distinction.

A fundamental distinction is between cross-classification and nesting. Thus in the simplest case we may have an array of observations, which we shall now denote by $Y_{a;j}$, in which the labelling of the repeat observations for each a is essentially arbitrary, i.e. $j = 1$ at $a = 1$ has no meaningful connection with $j = 1$ at $a = 2$. We say that the second suffix is *nested* within the first. By contrast we may have an arrangement Y_{ab}, which can be thought of as a row by column $A \times B$ array in which, say, the column labelling retains the same meaning for each row a and *vice versa*. We say that rows are *crossed* with columns.

Corresponding to these structures we may decompose the data vector into components. First, for the nested arrangement, we have that

$$Y_{a;j} = \bar{Y}_{..} + (\bar{Y}_{a.} - \bar{Y}_{..}) + (Y_{a;j} - \bar{Y}_{a.}). \tag{A.43}$$

This is to be contrasted with, for the cross-classification, the decomposition

$$Y_{ab} = \bar{Y}_{..} + (\bar{Y}_{a.} - \bar{Y}_{..}) + (\bar{Y}_{.b} - \bar{Y}_{..}) + (Y_{ab} - \bar{Y}_{a.} - \bar{Y}_{.b} + \bar{Y}_{..}). \tag{A.44}$$

As usual averaging over a suffix is denoted by a full-stop.

Now in these decompositions the terms on the right-hand sides, considered as defining vectors, are mutually orthogonal, leading to familiar decompositions of the sums of squares. Moreover there is a corresponding decomposition of the dimensionalities, or degrees of freedom, of the spaces in which these vectors lie, namely for the nested case

$$AJ = 1 + (A - 1) + A(J - 1)$$

ANALYSIS OF VARIANCE

and in the cross-classified case

$$AB = 1 + (A-1) + (B-1) + (A-1)(B-1).$$

Note that if we were mistakenly to treat the suffix j as if it were a meaningful basis for cross-classification we would be decomposing the third term in the analysis of the data vector, the sum of squares and the degrees of freedom into the third and fourth components of the crossed analysis. In general if a nested suffix is converted into a crossed suffix, variation within nested levels is typically converted into a main effect and one or more interaction terms.

The skeleton analysis of variance tables corresponding to these structures, with degrees of freedom, but without sums of squares, are given in Table A.2.

There are many possibilities for extension, still keeping to balanced arrangements. For example $Y_{abc;j}$ denotes an arrangement in which observations, perhaps corresponding to replicate determinations, are nested within each cell of an $A \times B \times C$ cross-classification, whereas observations $Y_{(a;j)bc}$ have within each level of the first classification a number of sublevels which are all then crossed with the levels of the other classifications. The skeleton analysis of variance tables for these two settings are given in Tables A.3 and A.4. Note that in the second analysis the final residual could be further decomposed.

These decompositions may initially be regarded as concise descriptions of the data structure. Note that no probabilistic considerations have been explicitly involved. In thinking about relatively

Table A.2 *Skeleton analysis of variance table for nested and crossed structures.*

| Nested | | Crossed | |
Source	D.f.	Source	D.f.
Mean	1	Mean	1
A-class (groups)	$A-1$	A-class (rows)	$A-1$
Within groups	$A(J-1)$	B-class (cols)	$B-1$
		$A \times B$	$(A-1)(B-1)$
Total	AJ		
		Total	AB

Table A.3 *Skeleton analysis of variance for $Y_{abc;j}$.*

Source	D.f.
Mean	1
A	$A-1$
B	$B-1$
C	$C-1$
$B \times C$	$(B-1)(C-1)$
$C \times A$	$(C-1)(A-1)$
$A \times B$	$(A-1)(B-1)$
$A \times B \times C$	$(A-1)(B-1)(C-1)$
Within cells	$ABC(J-1)$
Total	$ABCJ$

Table A.4 *Skeleton analysis of variance for $Y_{(a;j)bc}$.*

Source	D.f.
Mean	1
A	$A-1$
Within A	$A(J-1)$
B	$B-1$
C	$C-1$
$B \times C$	$(B-1)(C-1)$
$C \times A$	$(C-1)(A-1)$
$A \times B$	$(A-1)(B-1)$
$A \times B \times C$	$(A-1)(B-1)(C-1)$
Residual	$A(BC-1)(J-1)$
Total	$ABCJ$

complex arrangements it is often essential to establish which features are to be regarded as crossed and which as nested.

In terms of modelling it may then be useful to convert the data decomposition into a model in which the parameters associated with nested suffixes correspond to random effects, differing levels of nesting corresponding to different variance components. Also

ANALYSIS OF VARIANCE

it will sometimes be necessary to regard the highest order interactions as corresponding to random variables, in particular when one of the interacting factors represents grouping of the units made without specific subject-matter interpretation. Further all or most purely cross-classified suffixes correspond to parameters describing the systematic structure, either parameters of direct interest or characterizations of block and similar effects. For example, the model corresponding to the skeleton analysis of variance in Table A.4 is

$$Y_{(a;j)bc} = \tau_a^A + \eta_{j(a)} + \tau_b^B + \tau_c^C + \tau_{bc}^{BC} + \tau_{ab}^{AB} + \tau_{ac}^{AC} + \tau_{abc}^{ABC} + \epsilon_{j(abc)}.$$

For normal theory linear models with balanced data and a single level of nesting the resulting model is a standard linear model and the decomposition of the data and the sums of squares have a direct use in terms of standard least squares estimation and testing. With balanced data and several levels of nesting hierarchical error structures are involved.

Although we have restricted attention to balanced arrangements and normal error, the procedure outlined here suggests a systematic approach in more complex problems. We may have data unbalanced because of missing combinations or unequal replication. Further the simplest appropriate model may not be a normal theory linear model but, for example a linear logistic or linear probit model for binary response data. The following procedure may nevertheless be helpful.

First write down the formal analysis of variance table for the nearest balanced structure.

Next consider the corresponding normal theory linear model. If it has a single error term the corresponding linear model for unbalanced data can be analysed by least squares and the corresponding generalized linear model, for example for binary data, analysed by the method of maximum likelihood. Of course even in the normal theory case the lack of balance will mean that the sums of squares in the analysis of variance decomposition must be interpreted in light of the other terms in the model. If the model has multiple error terms the usual normal theory analysis uses the so-called residual maximum likelihood, or REML, for inference on the variance components. This involves constructing the marginal likelihood for the residuals after estimating the parameters in the mean. The corresponding analysis for generalized linear models will involve some special approximations.

The importance of these remarks lies in the need to have a systematic approach to developing models for complex data and for techniques of analysis when normal theory linear models are largely inapplicable.

A.4 More general models; maximum likelihood

We have, partly for simplicity of exposition and partly because of their immediate practical relevance, emphasized analyses for continuous responses based directly or indirectly on the normal theory linear model. Other types of response, in particular binary, ordinal or nominal, arise in many fields. Broadly speaking the use of standard designs, for example of the factorial type, will usually be sensible for such situations although formal optimality considerations will depend on the particular model appropriate for analysis, except perhaps locally near a null hypothesis; see Section 7.6.

For models other than the normal theory linear model, formal methods of analysis based on the log likelihood function or some modification thereof provide analyses serving essentially the same role as those available in the simpler situation. Thus confidence intervals based on the profile likelihood or one of its modifications are the preferred analogue of intervals based on the Student t distribution and likelihood ratio tests the analogue of tests for sets of parameters using the F distribution. We shall not review these further here.

A.5 Bibliographic notes

The method of least squares as applied to a linear model has a long history and an enormous literature. For the history, see Stigler (1986) and Hald (1998). Nearly but not quite all books on design of experiments and virtually all those on regression and analysis of variance and most books on general statistical theory have discussions of least squares and the associated distribution theory, the mathematical level and style of treatment varying substantially between books. The geometrical treatment of the distribution theory is perhaps implicit in comments of R. A. Fisher, and is certainly a natural development from his treatment of distributional problems. It was explicitly formulated by Bartlett (1933) and by Kruskal (1961) and in some detail in University of North Carolina lecture notes, unpublished so far as we know, by R. C. Bose.

Analysis of variance, first introduced, in fact in the context of a nonlinear model, by Fisher and Mackenzie (1923), is often presented as an outgrowth of linear model analysis and in particular either essentially as an algorithm for computing residual sums of squares or as a way of testing (usually uninteresting) null hypotheses. This is only one aspect of analysis of variance. An older and in our view more important role is in clarifying the structure of sets of data, especially relatively complicated mixtures of crossed and nested data. This indicates what contrasts can be estimated and the relevant basis for estimating error. From this viewpoint the analysis of variance table comes first, then the linear model not *vice versa*. See Nelder (1965a, b) for a systematic formulation and from a very much more mathematical viewpoint Speed (1987). Brien and Payne (1999) describe some further developments. The main systematic treatment of the theory of analysis of variance remains the book of Scheffé (1959).

The method of fitting in stages is implicit in Gauss's treatment; the discussion in Draper and Smith (1998, Chapter 2) is helpful. The connection with analysis of covariance goes back to the introduction of that technique; for a broad review of analysis of covariance, see Cox and McCullagh (1982). The relation between least squares theory and asymptotic theory is discussed by van der Vaart (1998).

The approach to fitting models of complex structure to generalized linear models is discussed in detail in McCullagh and Nelder (1989).

For techniques for assessing the adequacy of linear models, see Atkinson (1985) and Cook and Weisberg (1982).

For discussion of floating parameters and pseudo-variances, see Ridout (1989), Easton et al. (1991), Reeves (1991) and Firth and Menezes (2000).

A.6 Further results and exercises

1. Show that the matrices $X(X^TX)^{-1}X^T$ and $I - X(X^TX)^{-1}X^T$ are idempotent, i.e. equal to their own squares and give the geometrical interpretation. The first is sometimes called the hat matrix because of its role in forming the vector of fitted values $\hat{Y}$.

2. If the covariance matrix of Y is $V\sigma^2$, where V is a known positive definite matrix, show that the geometrical interpretation

of Section A.2.3 is preserved when norm and scalar product are defined by
$$\|Y\|^2 = Y^T V^{-1} Y, \quad (Y_1, Y_2) = Y_1^T V^{-1} Y_2.$$
Show that this leads to the generalized least squares estimating equation
$$X^T V^{-1}(Y - X\hat{\theta}) = 0.$$
Explain why these are appropriate definitions statistically and geometrically.

3. Verify by direct calculation that in the least squares analysis of a completely randomized design essentially equivalent answers are obtained whatever admissible constraint is imposed on the treatment parameters.

4. Denote the jth diagonal element of $X(X^T X)^{-1} X^T$ by h_j, called the *leverage* of the corresponding response value. Show that $0 < h_j < 1$ and that $\Sigma h_j = q$, where q is the rank of $X^T X$. Show that the variance of the corresponding component of the residual vector, $Y_{\text{res},j}$ is $\sigma^2(1 - h_j)$, leading to the definition of the jth standardized residual as $r_j = Y_{\text{res},j}/\{s\sqrt{(1 - h_j)}\}$, where s^2 is the residual mean square. Show that the difference between Y_j and the predicted mean for Y_j after omitting the jth value from the analysis, $x_j^T \hat{\beta}_{(j)}$, divided by the estimated standard error of the difference (again omitting Y_j from the analysis) is
$$r_j(n - q - 1)^{1/2}/(n - q - r_j^2)^{1/2}$$
and may be helpful for examining for possible outliers. For some purposes it is more relevant to consider the influence of specific observations on particular parameters of interest. See Atkinson (1985) and Cook and Weisberg (1982, Ch. 2).

5. Suggest a procedure for detecting a single defective observation in a single Latin square design. Test the procedure by simulation on a 4×4 and a 8×8 square.

6. Develop the intra-block analysis of a balanced incomplete block design via the method of fitting parameters in stages.

7. Consider a $v \times v$ Latin square design with a single baseline variable z. It is required to fit the standard Latin square model augmented by linear regression on z. Show by the method of fitting parameters in stages that this is achieved by the following

FURTHER RESULTS AND EXERCISES 247

construction. Write down the standard Latin square analysis of variance table for the response Y and the analogous forms for the sum of products of Y with z and the sum of squares of z.

Let R_{YY}, R_{zY}, R_{zz} and T_{YY}, T_{zY}, T_{zz} denote respectively the residual and treatment lines for these three analysis of variance tables. Then

(a) the estimated regression coefficient of Y on z is R_{zY}/R_{zz};
(b) the residual mean square of Y in the full analysis is $(R_{YY} - R_{zY}^2/R_{zz})/(v^2 - 3v + 1)$;
(c) the treatment effects are estimated by applying an adjustment proportional to R_{zY}/R_{zz} to the simple treatment effects estimated from Y in an analysis ignoring z;
(d) the adjustment is uncorrelated with the simple unadjusted effects so that the standard error of an adjusted estimate is easily calculated;
(e) an F test for the nullity of all treatment effects is obtained by comparing with the above residual mean square the sum of squares with $v - 1$ degrees of freedom
$$T_{YY} + R_{YY} - (T_{zY} + R_{zY})^2/(T_{zz} + R_{zz}).$$

How would treatment by z interaction be tested?

8. Recall the definition of a sum of squares with one degree of freedom associated with a contrast $l^T Y = \Sigma l_j Y_j$, with $\Sigma l_j = 0$, as $(l^T Y)^2/(l^T l)$. Show that in the context of the linear model $E(Y) = X\theta$, the contrast is a function only of the residual vector if and only if $l^T X = 0$. In this case show that under the normal theory assumption l can be taken to be any function of the fitted values $\hat{Y}$ and the distribution of the contrast will remain σ^2 times chi-squared with one degree of freedom.

For a randomized block design the contrast $\Sigma l_{js} Y_{js} = \Sigma l_{js}(Y_{js} - \hat{Y}_{js})$ with $l_{js} = (\bar{Y}_{j.} - \bar{Y}_{..})(\bar{Y}_{.s} - \bar{Y}_{..})$ was suggested by Tukey (1949) as a means of checking deviations from additivity of the form
$$E(Y_{js}) = \mu + \tau_j + \beta_s + \gamma(\tau_j \beta_s).$$

APPENDIX B
Some algebra

B.1 Introduction

Some specialized aspects of the design of experiments, especially the construction of arrangements with special properties, have links with problems of general combinatorial interest. This is not a topic we have emphasized in the book, but in this Appendix we review in outline some of the algebra involved. The discussion is in a number of sections which can be read largely independently. One objective of this Appendix is to introduce some key algebraic ideas needed to approach some of the more specialized literature.

B.2 Group theory

B.2.1 Definition

A *group* is a set $\mathcal{G}$ of elements $\{a, b, \ldots\}$ and a rule of combination such that

G1 for each ordered pair $a, b \in \mathcal{G}$, there is defined a unique element $ab \in \mathcal{G}$;

G2 $(ab)c = a(bc)$;

G3 there exists $e \in \mathcal{G}$, such that $ea = a$, for all $a \in \mathcal{G}$;

G4 for each $a \in \mathcal{G}$, there exists $a^{-1} \in \mathcal{G}$ such that $a^{-1}a = e$.

Many properties follow directly from **G1–G4**. Thus $ae = ea$, $aa^{-1} = a^{-1}a = e$, and e, a^{-1} are unique. Also $ax = ay$ implies $x = y$.

If $ab = ba$ for all $a, b \in \mathcal{G}$, then $\mathcal{G}$ is called *commutative* (or Abelian). If it has a finite number n of elements we call it a *finite group of order* n. All the groups considered in the present context are finite.

A subset of elements of $\mathcal{G}$ forming a group under the law of combination of $\mathcal{G}$ is called a *subgroup* of $\mathcal{G}$.

Starting with a subgroup $\mathcal{S}$ of order r we can generate the group $\mathcal{G}$ by multiplying the elements of $\mathcal{S}$, say on the left, by new elements to form what are called the *cosets* of $\mathcal{S}$. This construction is used repeatedly in the theory of fractional replication and confounding.

We could equally denote the law of combination by $+$, but by convention we restrict $+$ to commutative groups and then denote the identity by 0.

Examples

1. all integers under addition (infinite group)
2. the cyclic group of order n: $\mathcal{C}_n(a)$. This is the set

$$\{1, a, a^2, \ldots, a^{n-1}\} \tag{B.1}$$

with the rule $a^r a^s = a^t$ where $t = (r+s) \bmod n$. Alternatively this can be written as the additive group of least positive residues mod n, G_n^+, $G_n^+ = \{0, 1, \ldots, n-1\}$, with the rule $r + s = t$, where $t = (r+s) \bmod n$. Clearly $\mathcal{C}_n$ and G_n^+ are essentially the same group. They are said to be *isomorphic*; the elements of the two groups can be placed in 1-1 correspondence in a way preserving the group operation.

B.2.2 Prime power commutative groups

Let p be prime. Build up groups from $\mathcal{C}_p(a)$, $\mathcal{C}_p(b)$ as follows:

$$\begin{array}{ccccc} 1 & a & a^2 & \ldots & a^{p-1} \\ b & ab & a^2 b & \ldots & a^{p-1} b \\ \vdots & \vdots & \vdots & & \vdots \\ b^{p-1} & ab^{p-1} & a^2 b^{p-1} & \ldots & a^{p-1} b^{p-1} \end{array} \tag{B.2}$$

This set of p^2 symbols forms a commutative group if we define $(a^i b^j)(a^k b^l) = a^{i+k} b^{j+l}$, reducing mod p where necessary. We call this group $\mathcal{C}_p(a, b)$. Similarly, with the symbols $a, b, \ldots, d$ we define $\mathcal{C}_p(a, b, \ldots, d)$ to be the set of all powers $a^i b^j \ldots d^k$, with indices between $0, \ldots, p-1$. The group is called a *prime power commutative group* of order p^m and $a, b, \ldots, d$ are called the *generators*.

Properties

1. A group is generated equally by any set of m independent elements.
2. Any subgroup is of order $p, p^2, \ldots$ and is generated by a suitable set of elements.

In the group $C_p(a,b)$ enumerated above, the first line is a subgroup. The remaining lines are obtained by repeated multiplication by fixed elements and are thus the cosets of the subgroup.

B.2.3 Permutation groups

We now consider a different kind of finite group. Let each element of the group denote a permutation of the positions

$$\{1, 2, \ldots, n\}.$$

For example, with $n = 4$ the elements a and b might produce

$$a: \quad 2, 4, 3, 1;$$
$$b: \quad 4, 3, 2, 1.$$

We define ab by composition, i.e. by applying first b then a to give in the above example

$$1, 3, 4, 2;$$

note that ba here gives

$$3, 1, 2, 4$$

showing that in general composition of permutations is not commutative. A group of permutations is a set of permutations such that if a and b are in the set so too is ab; the unit element leaves all positions unchanged and we require also the inclusion with every a, its inverse a^{-1} defined by $aa^{-1} = e$, i.e. by restoring the original positions.

The simplest such group is the set of all possible permutations, called the *symmetric group* of order n, denoted by $\mathcal{S}_n$; it has $n!$ elements.

A group of transformations is called *transitive* if it contains at least one permutation sending position i into position j for all i, j. It is called *doubly transitive* if it contains a permutation sending any ordered pair $i, j; i \neq j$ into any other ordered pair $k, l; k \neq l$. It can be shown that if $i \neq j \neq k \neq l$ then the number of permutations with the required property is the same for all i, j, k, l.

An important construction of a group of permutations is as follows. Divide the positions into b blocks each of k positions. Consider a group formed as follows. Take a permutation from the symmetric group $\mathcal{S}_b$ to permute the blocks. Then take permutations from b separate symmetric groups $\mathcal{S}_k$ to permute positions within blocks. The group formed by composing these permutations is called a *wreath product*.

B.2.4 Application to randomization theory

The most direct way of thinking about the randomization of a design is to consider the experimental units as given, labelled $1, \ldots, n$, say and then a particular pattern of treatment allocation to be chosen at random out of some suitable set of arrangements. In this sense the units are fixed and the treatments randomized to the units. It is equivalent, however, to suppose that a treatment pattern is fixed and then the units allocated at random to that pattern.

For example consider an experiment with two treatments, T and C and six units $1, \ldots, 6$, a completely randomized design with equal replication being used. We may start with the design

$$T, T, T, C, C, C.$$

Next apply the symmetric group $\mathcal{S}_6$ of 6! permutations to the initial order $1, 2, 3, 4, 5, 6$ of the six experimental units. This generates the set

$$1, 2, 3, 4, 5, 6$$
$$1, 2, 3, 4, 6, 5$$
$$\vdots$$
$$6, 5, 4, 3, 2, 1$$

Then we choose one permutation at random out of that set as the specification of the design. In some respects this is a clumsy construction but it has the advantage of making it clear that because the set of possible designs is invariant under any permutation in $\mathcal{S}_6$ so too must be the properties of the randomization distribution.

If instead we had used the matched pair design based on the pairs $(1, 2), (3, 4), (5, 6)$ the initial design would have been

$$T, C, T, C, T, C.$$

The permutations would either have interchanged units within a pair or interchanged units within a pair and pairs as a whole.

The first possibility gives

$$1, 2, 3, 4, 5, 6$$
$$1, 2, 3, 4, 6, 5$$
$$1, 2, 4, 3, 5, 6$$
$$1, 2, 4, 3, 6, 5$$

GROUP THEORY

$$2, 1, 3, 4, 5, 6$$
$$2, 1, 3, 4, 6, 5$$
$$2, 1, 4, 3, 5, 6$$
$$2, 1, 4, 3, 6, 5$$

The second possibility, which would become especially relevant if a second set of treatments was to be imposed at the pair level, would involve also interchanging pairs to give, for example

$$3, 4, 1, 2, 5, 6$$
$$3, 4, 2, 1, 5, 6$$

etc. In the first case, the set of designs is invariant under the permutation group consisting of all possible transpositions of pairs. In the second a larger group is involved, in fact the wreath product as defined above. Again it follows that the randomization distributions involved are invariant under the appropriate group.

When the second moment theory of randomization is considered we are concerned only with the properties of linear and quadratic functions of the data. The arguments used in Sections 3.3 and 3.4 use invariance to simplify the randomization expectations involved. If in the present formulation the set of designs is invariant under a group $\mathcal{G}$ of permutations so too are all expectations involving the unit constants ξ in the notation of Chapters 2 and 3.

There are essentially two uses of this formulation. One is in connection with complex designs where randomization has been used but it is not clear to what extent a randomization-based analysis is valid. Then clarification of the group of permutations that leaves the design invariant can resolve the issue.

Secondly, even in simple designs where second moment properties are considered, the maximal group of permutations may not be required; the key property is some version of double transitivity because the focus of interest is the randomization expectation of quadratic forms. This leads to the notion of restricted randomization; it may be possible to label certain arrangements generated by the "full" group as objectionable and to find a restricted group of permutations having the right double transitivity properties to justify the standard analysis but excluding the objectionable arrangements. The explicit use of permutation groups is unnecessary for the relatively simple designs considered in this book but is essential for more complex possibilities.

B.3 Galois fields

B.3.1 Definition

The most important algebraic systems involving two operations, by convention represented as addition and multiplication, are known as *fields* and we give a brief introduction to their properties.

A field is a set $\mathcal{F}$ of elements $\{a, b, \ldots\}$ such that for any pair $a, b \in \mathcal{F}$, there are defined unique $a + b, a \cdot b \in \mathcal{F}$ such that

F1 Under $+$, $\mathcal{F}$ is an additive group with identity 0;

F2 Under $\cdot$, all elements of $\mathcal{F}$ except 0 form a commutative group;

F3 $a \cdot (b + c) = a \cdot b + a \cdot c$.

Various properties follow from the axioms. In particular we have cancellation laws: $a + b = a + c$ implies $b = c$; $a \cdot b = a \cdot c$ and $a \neq 0$, imply $b = c$. **F2** implies the existence of a unit element.

If $\mathcal{F}$ contains a finite number n of elements it is called a Galois field of order n.

The key facts about Galois fields are:

1. Galois fields exist if and only if n is a prime power p^m;
2. any two fields of order n are isomorphic, so that there exists essentially only one Galois field of order p^m, which we denote $\mathrm{GF}(p^m)$.

Fields of prime order, $\mathrm{GF}(p)$, may be defined as consisting of $\{0, 1, \ldots, p-1\}$, defining addition and multiplication mod p. This construction satisfies **F1** for all p, but **F2** only for prime p.

We use $\mathrm{GF}(p)$ to construct finite fields of prime power order, $\mathrm{GF}(p^m)$. These consist of all polynomials

$$a_0 + a_1 x + \ldots + a_{m-1} x^{m-1},$$

with $a_i \in \mathrm{GF}(p)$. Obviously there are p^m such expressions. Addition is defined as ordinary addition with reduction of the coefficients mod p. To define multiplication, we use an irreducible polynomial $P(x)$ of degree m, and with coefficients in $\mathrm{GF}(p)$. That is we take

$$P(x) = \alpha_0 + \alpha_1 x + \ldots + \alpha_m x^m \qquad (\alpha_m \neq 0), \qquad (B.3)$$

where $P(x)$ is not a product, reducing mod p, of polynomials of lower degree. Such a $P(x)$ always exists. The product of two elements in $\mathrm{GF}(p^m)$ is the remainder of their ordinary product after division by $P(x)$ and reduction of the coefficients mod p. It can be shown that this defines a field.

GALOIS FIELDS   255

Example. For $\mathrm{GF}(2^2)$ the elements are $\{0, 1, x, x+1\}$. An irreducible polynomial is $P(x) = x^2 + x + 1$; note that $P(x)$ is not x^2 or $x \cdot (x+1)$ or $(x+1)^2 = x^2 + 1$. Then for example $x \cdot x$ is

$$x^2 = (x^2 + x + 1) - (x+1) = -(x+1) = x+1, \qquad (\text{B.4})$$

after division by $P(x)$ and because $-1 = 1$.

The field $\mathrm{GF}(p^m)$ can alternatively be constructed using a *power cycle* from a *primitive element* in which the powers $x^1, \ldots, x^{p^m-1}$ are identified with each nonzero element of the field. In the Example above $x^1 = x, x^2 = x+1, x^3 = 1$. The powers $x^1, \ldots, x^{p^m-1}$ contain each nonzero element of the field just once. The power cycle can be used to work back to the multiplication table; e.g. $x \cdot (x+1) = xx^2 = 1$.

We define a nonzero member a of the field to be a *quadratic residue* if it is the square of another member of the field. It can be seen that quadratic residues are even powers of a primitive element of the field and that therefore the number of nonzero nonquadratic residues is the same as the number of quadratic residues. That is, if we define

$$\chi(a) = \begin{cases} 1 & a \text{ is a quadratic residue} \\ -1 & a \neq 0 \text{ and is not a quadratic residue} \\ 0 & a = 0, \end{cases} \qquad (\text{B.5})$$

then

$$\Sigma \chi(a) = 0, \quad \chi(a)\chi(b) = \chi(ab), \qquad (\text{B.6})$$
$$\Sigma_j \chi(j - i_1)\chi(j - i_2) = -1, \qquad (\text{B.7})$$

the last, a quasi-orthogonality relation, being useful in connection with the construction of Hadamard matrices.

B.3.2 Orthogonal sets of Latin squares

We now sketch the application of Galois fields to orthogonal Latin squares, adopting a rather more formal approach than that sketched in Section 4.1.3.

A set of $n \times n$ Latin squares such that any pair are orthogonal is called an *orthogonal set* and an orthogonal set of $n-1$ $n \times n$ Latin squares is called a *complete orthogonal set*.

The central result is that whenever n is a prime power p^m, a complete orthogonal set exists. To see this, number the rows, columns and letters by the elements of $\mathrm{GF}(p^m)$; $u_0 = 0, u_1 =$

$1, u_2, \ldots, u_{n-1}$. For each $\lambda = 1, \ldots, n-1$ define a Latin square L_λ by the rule: in row u_x, column u_y, put letter $u_\lambda u_x + u_y$. Symbolically

$$L_\lambda : \{u_x, u_y : u_\lambda u_x + u_y\}. \tag{B.8}$$

Then these are a complete orthogonal set. For

1. L_λ is a Latin square;
2. if $\lambda \neq \lambda'$, L_λ and $L_{\lambda'}$ are orthogonal.

To prove 1., note that if the same letter occurs in row u_x and columns u_y and $u_{y'}$, then

$$u_\lambda u_x + u_y = u_\lambda u_x + u_{y'}. \tag{B.9}$$

This implies that $u_y = u_{y'}$, because of the cancellation law in the additive group $\mathrm{GF}(p^m)$.

Similarly if the same letter occurs in rows $u_x, u_{x'}$, and in column u_y, then

$$\begin{aligned} u_\lambda u_x + u_y &= u_\lambda u_{x'} + u_y, \\ u_\lambda u_x &= u_\lambda u_{x'}, \\ u_x &= u_{x'} \end{aligned}$$

using both addition and multiplication cancellation laws.

To prove 2., suppose that row u_x, column u_y contain the same pair of letters as row $u_{x'}$ and column $u_{y'}$. Then

$$\begin{aligned} u_\lambda u_x + u_y &= u_\lambda u_{x'} + u_{y'}, \\ u_{\lambda'} u_x + u_y &= u_{\lambda'} u_{x'} + u_{y'}. \end{aligned}$$

Therefore $(u_\lambda - u_{\lambda'}) u_x = (u_\lambda - u_{\lambda'}) u_{x'}$. Thus $u_x = u_{x'}$, since $u_\lambda - u_{\lambda'} \neq 0$. Similarly $u_y = u_{y'}$.

From this it follows that any square of the set can be derived from any other, in particular from L_1, by a permutation of rows. For if and only if $u_\lambda u_x = u_{\lambda'} u_{x'}$, then the u_x row of L_λ is identical with the $u_{x'}$ row of $L_{\lambda'}$. Further the last equation has a unique solution for $u_{x'}$, so that each row of L_λ occurs just once in $L_{\lambda'}$. (This result is not true for all complete orthogonal sets.)

A second consequence is that L_1 is the addition table of $\mathrm{GF}(p^m)$. For it is given by the rule $\{u_x, u_y; u_x + u_y\}$. An example of two orthogonal 5×5 Latin squares is given in Table B.1.

Let $N(n)$ be the maximum possible number of squares in an orthogonal set of $n \times n$ squares. We have shown above that if $n = p^m$, $N(n) = n - 1$. This can be extended to show that if

GALOIS FIELDS

Table B.1 *The construction of two orthogonal 5×5 Latin squares.*

		Column				
		u_0	u_1	u_2	u_3	u_4
Row	$u_0 = 0$	0	1	2	3	4
	$u_1 = 1$	1	2	3	4	0
	$u_2 = 2$	2	3	4	0	1
	$u_3 = 3$	3	4	0	1	2
	$u_4 = 4$	4	0	1	2	3
			L_1			

		u_0	u_1	u_2	u_3	u_4
Row	$u_0 = 0$	0	1	2	3	4
	$u_1 = 1$	2	3	4	0	1
	$u_2 = 2$	4	0	1	2	3
	$u_3 = 3$	1	2	3	4	0
	$u_4 = 4$	3	4	0	1	2
			L_2			

$n = p_1^{m_1} p_2^{m_2} \ldots$ $(p_1 \neq p_2 \neq \ldots)$ and if $r_n = \min(p_1^{m_1}, p_2^{m_2}, \ldots,)$, then $N(n) \geq r_n - 1$. Thus if $n = 12$, $r_n = \min(2^2, 3) = 3$ and there exists at least $3 - 1 = 2$ orthogonal 12×12 squares. Fisher and Yates, and others, have shown that $N(6) = 1$, i.e. there is not even a Graeco-Latin square of size 6.

A longstanding conjecture of Euler was that $N(n) = r_n - 1$, which would have implied that no Graeco-Latin square exists when $n = 2 \mod 4$, and in particular that no 10×10 Graeco-Latin square exists. A pair of orthogonal 10×10 Latin squares was constructed in 1960, and it is now known that $N(n) > 1$, $n > 6$, so that Graeco-Latin squares exist except when $n = 6$. Some bounds for $N(n)$ are known, and $N(n) \to \infty$ as $n \to \infty$.

Notes

1. The full axioms for a field are not used in the above construction. It would be enough to have a linear associative algebra. This fact does lead to systems of squares essentially different from $L_1, \ldots, L_n$, but not to a solution when $n \neq p^m$.

2. Orthogonal partitions of Latin squares can be constructed: the following is an example derived from a 4×4 Graeco-Latin square.

Bγ	Dα	Cβ	Aδ	BII	DI	CI	AII
Cδ	Aβ	Bα	Dγ	CII	AI	BI	DII
Aα	Cγ	Dδ	Bβ	AI	CII	DII	BI
Dβ	Bδ	Aγ	Cα	DI	BII	AII	CI

The symbols I, II each occur twice in each row and column and twice in common with each letter. Orthogonal partitions exist for 6 × 6 squares.

3. Combinatorial properties of Latin squares are unaffected by changes between rows, columns and letters.

B.4 Finite geometries

Closely associated with Galois fields are systems of finite numbers of "points" that with suitable definitions satisfy axioms of either Euclidean or projective geometry and are therefore reasonably called finite geometries. In an abstract approach we start with a system consisting of a finite number of *points* and a collection of *lines*, each line consisting of a set of points said to be collinear. Such a system is called a *finite projective geometry* $\mathrm{PG}(k, p^m)$ if it obeys the following axioms:

1. there is just one line through any pair of points

2. if points A, B, C are not collinear and if a line l contains a point D on the line AB and a point E on the line BC, then it contains a point F on the line CA

3. if points are called 0-spaces and lines 1-spaces and if q-spaces are defined inductively, for example by defining 2-spaces as the set of points collinear with points on two given distinct lines, then if $q < k$ not all points lie in the same q space and there is no $k+1$ space

4. there are $(p^m)^k + p^m + 1$ distinct points.

It can be shown that such a system is isomorphic with the following construction. Let a_j denote elements of $\mathrm{GF}(p^m)$. Then a point is identified by a set of homogeneous coordinates

$$(a_0, a_1, \ldots, a_k),$$

FINITE GEOMETRIES

not all zero. By the term homogeneous coordinates is meant that all sets of coordinates $(aa_0, aa_1, \ldots, aa_k)$ for a any nonzero element of the field denote the same point. The line joining two points

$$(a_0, a_1, \ldots, a_k), \quad (b_0, b_1, \ldots, b_k)$$

contains all points with coordinates

$$(\lambda a_0 + \mu b_0, \lambda a_1 + \mu b_1, \ldots, \lambda a_k + \mu b_k), \tag{B.10}$$

where λ, μ are elements of the field.

The axioms defining a field can be used to show that the requirements of a finite projective geometry are satisfied.

Now in "ordinary" geometry, Euclidean geometry is obtained from a corresponding projective geometry by deleting points at infinity. A similar construction is possible here. Take all those points with $a_0 \neq 0$; without loss of generality we can then set a_0 to the unit element of the field and take the remaining points as defined by a unique set of k coordinates, $a_1, \ldots, a_k$, say. This system is called a *finite Euclidean geometry*, $EG(k, p^m)$. In effect the set of points with $a_0 = 0$ plays the role of a point at infinity.

Many of the features of "ordinary" geometry, for example a duality principle in which $k-1$ subspaces correspond to points and $k-2$ subspaces to lines can be mimicked in these systems.

In particular if $m_1 < m$ and we define a subfield $GF(p^{m_1})$ contained in $GF(p^m)$ we can derive a proper subgeometry $PG(k, p^{m_1})$ within $PG(k, p^m)$ by using only elements of the subfield.

As an example we consider $PG(2, 2)$ contained within $PG(2, 2^2)$. We start with the elements of the Galois field labelled $\{0, 1, x, x+1\}$ as above. The full system has 21 points with homogeneous coordinates as follows:

A	B	C	D	E
$0, 0, 1$	$0, 1, 0$	$0, 1, 1$	$0, 1, x$	$0, 1, x+1$
F	G	H	I	J
$1, 0, 0$	$1, 0, 1$	$1, 0, x$	$1, 0, x+1$	$1, 1, 0$
K	L	M	N	O
$1, 1, 1$	$1, 1, x$	$1, 1, x+1$	$1, x, 0$	$1, x, 1$
P	Q	R	S	T
$1, x, x$	$1, x, x+1$	$1, x+1, 0$	$1, x+1, 1$	$1, x+1, x$
U				
$1, x+1, x+1$				

The lines are formed from linear combinations of coordinates. Thus on the line AB are also the points $0, \mu A + \lambda B$ for all choices of λ, μ from the nonzero elements of $\mathrm{GF}(2^2)$. This leads to the line containing just the points A, B, C, D, E.

The subgeometry $\mathrm{PG}(2, 2)$ is formed from the points with coordinates 00, 01, 10, 11 formed from the elements 0, 1 of $\mathrm{GF}(2)$. These are the points A, B, C, F, G, J, K in the above specification and when these are arranged in a rectangle with associated lines as columns we obtain the balanced incomplete block design with seven points (treatments) arranged in seven lines (blocks), with three points on each line, each pair of treatments occurring in the same line just once:

$$\begin{array}{ccccccc} A & B & F & C & J & K & G \\ B & F & C & J & K & G & A \\ C & J & K & G & A & B & F \end{array} \quad (\text{B.11})$$

A Euclidean geometry is formed from points $F, \ldots, U$ specifying each point by the second and third coordinate, for example M as $(1, x + 1)$.

B.5 Difference sets

A very convenient way of generating block, and more generally row by column, designs is by development from an initial block by repeated addition of 1. That is, if there are v treatments labelled $0, 1, \ldots, v - 1$ we define an initial block and then produce more blocks by successive addition of 1 and reduction mod v. For example, if $v = 7$ and we start with the initial block 1, 2, 4 then with the successive blocks, namely 2, 3, 5; 3, 4, 6; 4, 5, 0; 5, 6, 1; 6, 0, 2; 0, 1, 3, we have a balanced incomplete block design with $v = b = 7, r = k = 3, \lambda = 1$, different from that in (B.11).

The key to this construction is that in the initial block the differences between pairs of entries are $3 - 2 = 1$, $2 - 3 = -1 = 6$, $5 - 2 = 3$, $2 - 5 = 4$, $5 - 3 = 2$, $3 - 5 = 5$, so that each possible difference occurs just once. This implies that in the whole design each pair of treatments occurs together just once.

There are connections between difference sets and Abelian groups and also with Galois fields. Thus it can be shown that for $v = p^m = 4q + 1$ there are two starting blocks with the desired properties, namely the set of nonzero quadratic residues of $\mathrm{GF}(p^m)$ and the set of nonzero nonquadratic residues. If $v = p^m = 4q - 1$ we take the nonzero quadratic residues.

B.6 Hadamard matrices

An $n \times n$ square matrix L is orthogonal by definition if

$$L^T L = I, \qquad (B.12)$$

where I is the $n \times n$ identity matrix. The matrix L may be called orthogonal in the extended sense if

$$L^T L = D, \qquad (B.13)$$

where D is a diagonal matrix with strictly positive elements. Such a matrix is formed from mutually orthogonal columns which are, however, not in general scaled to have unit norm. The columns of such a matrix can always be rescaled to produce an orthogonal matrix.

An $n \times n$ matrix H is called a *Hadamard matrix* if its first column consists only of elements $+1$, if its remaining elements are $+1$ or -1 and if it is orthogonal in the extended sense.

For such a matrix to exist n must be a multiple of 4. It has been shown that such matrices indeed exist for all multiples of 4 up to and including 424.

If for a prime p, $4t = p + 1$ we may define a matrix by

$$h_{i0} = h_{0j} = 1, h_{ii} = -1, h_{ij} = \chi(j - i) \qquad (B.14)$$

and the required orthogonality property follows from those of the quadratic residue. If $p^m = 4t - 1$ we proceed similarly labelling the rows and columns by the elements of $\mathrm{GF}(p^m)$. The size of a Hadamard matrix, H, can always be doubled by the construction

$$\begin{pmatrix} H & H \\ H & -H \end{pmatrix}. \qquad (B.15)$$

The following is a 8×8 Hadamard matrix:

$$\begin{pmatrix} 1 & 1 & 1 & 1 & 1 & 1 & 1 & 1 \\ 1 & 1 & 1 & 1 & -1 & -1 & -1 & -1 \\ 1 & 1 & -1 & -1 & 1 & 1 & -1 & -1 \\ 1 & 1 & -1 & -1 & -1 & -1 & 1 & 1 \\ 1 & -1 & 1 & -1 & 1 & -1 & 1 & -1 \\ 1 & -1 & 1 & -1 & -1 & 1 & -1 & 1 \\ 1 & -1 & -1 & 1 & 1 & -1 & -1 & 1 \\ 1 & -1 & -1 & 1 & -1 & 1 & 1 & -1 \end{pmatrix}. \qquad (B.16)$$

This is used to define the treatment contrasts in a 2^3 factorial in Section 5.5 and a saturated main effect plan for a 2^7 factorial in Section 6.3.

B.7 Orthogonal arrays

In Section 6.3 we briefly described orthogonal arrays, which from one point of view are generalizations of fractional factorial designs. The construction of orthogonal arrays is based on the algebra of finite fields. Orthogonal arrays can be constructed from Hadamard matrices, as illustrated in Section 6.3, and can also be constructed from Galois fields, from difference schemes, and from sets of orthogonal Latin squares.

A symmetric orthogonal array of size n with k columns has s symbols in each column, and has strength r if every $n \times r$ subarray contains each r-tuple of symbols the same number of times. Suppose $s = p^m$ is a prime power. Then an orthogonal array with $n = s^l$ rows and $(s^l - 1)/(s - 1)$ columns that has strength 2 can be constructed as follows: form an $l \times (s^l - 1)/(s - 1)$ matrix whose columns are all nonzero l-tuples from $GF(s)$ in which the first nonzero element is 1. All linear combinations of the rows of this generator matrix form an orthogonal array of the required size. This is known as the *Rao-Hamming* construction.

For example, with $s = 2$ and $l = 3$ and generator matrix

$$\begin{pmatrix} 1 & 0 & 0 & 1 & 1 & 0 & 1 \\ 0 & 1 & 0 & 1 & 0 & 1 & 1 \\ 0 & 0 & 1 & 0 & 1 & 1 & 1 \end{pmatrix}$$

CODING THEORY

we obtain the 8×7 orthogonal array of strength 2:

$$\begin{pmatrix} 0 & 0 & 0 & 0 & 0 & 0 & 0 \\ 1 & 0 & 0 & 1 & 1 & 0 & 1 \\ 0 & 1 & 0 & 1 & 0 & 1 & 1 \\ 0 & 0 & 1 & 0 & 1 & 1 & 1 \\ 1 & 1 & 0 & 0 & 1 & 1 & 0 \\ 1 & 0 & 1 & 1 & 0 & 1 & 0 \\ 0 & 1 & 1 & 1 & 1 & 0 & 0 \\ 1 & 1 & 1 & 0 & 0 & 0 & 1 \end{pmatrix}.$$

Orthogonal arrays can also be constructed from error-correcting codes, by associating to each codeword a row of an orthogonal array. In the next section we illustrate the construction of codes from some of the designs considered in this book.

B.8 Coding theory

The combinatorial considerations involved in the design of experiments, in particular those associated with orthogonal Latin squares and balanced incomplete block designs, have other applications, notably to the theory of error-detecting and error-correcting codes.

For example, suppose that $q \geq 3$, $q \neq 6$ so that a $q \times q$ Graeco-Latin square exists. Then we may use an alphabet of q letters $0, 1, \ldots, q-1$ to assign each of q^2 codewords a code of four symbols by labelling the codewords (i, j) for $i, j = 0, \ldots, q-1$ and then assigning codeword (i, j) the code

$$ij\alpha_{ij}\beta_{ij}, \qquad (B.17)$$

where α_{ij} and β_{ij} refer to the Latin and Greek letters in row i and column j translated onto $0, \ldots, q-1$ in the obvious way.

For example with $q = 3$ we obtain the following:

Codeword	00	01	02	10	11
Code	0000	0111	0222	1012	1120

Codeword		12	20	21	22
Code		1201	2021	2102	2210

It can be checked in this example, and indeed in general from the properties of the Graeco-Latin square, that the codes for any two codewords differ by at least three symbols. This implies that two errors in coding can be detected and one error corrected, the last by moving to the codeword nearest to the transmitted word. In the same way if $q = p^m$, we can by labelling the codewords via the elements of $\mathrm{GF}(p^m)$ and using the complete set of mutually orthogonal $q \times q$ Latin squares obtain a coding of q^2 codewords with $q + 1$ symbols per codeword and with very strong error detecting and error correcting properties.

Symmetrical balanced incomplete block designs with $b = v$ can be used to derive binary codes in a rather different way. Add to the incidence matrix of the design a row of 0's and below this the matrix with 0's and 1's interchanged, thus producing a $2v + 2$ by v matrix coding $2v + 2$ codewords with v symbols per codeword. Thus with $b = v = 7$, $r = k = 3$, $\lambda = 1$, sixteen codewords are each assigned seven binary symbols. Again any two codewords differ by at least three symbols and the error-detecting and error-correcting properties are as before.

B.9 Bibliographic notes

Restricted randomization was introduced by Yates (1951a, b) to deal with a particular practical problem arising with a quasi-Latin square, i.e. a factorial experiment with double confounding in square form. The group theory justification is due to Grundy and Healy (1950). The method was rediscovered in a much simpler context by Youden (1956). See the general discussion by Bailey and Rowley (1987).

Galois fields were introduced into the study of sets of Latin squares by Bose (1938).

A beautiful account of finite groups and fields is by Carmichael (1937). Street and Street (1981) give a wide-ranging account of combinatorial problems connected with experimental design.

John and Williams (1995) give an extensive discussion of designs formed by cyclic generation.

For the first account of a 10×10 Graeco-Latin square, see Bose, Shrikhande and Parker (1960). For orthogonal partitions of Latin squares, see Finney (1945a).

For an introduction to coding theory establishing connections with experimental design, see Hill (1986). The existence and construction of orthogonal arrays with a view to their statistical applications is given by Hedayat, Sloane and Stufken (1999). The use of error-correcting codes to construct orthogonal arrays is the subject

FURTHER RESULTS AND EXERCISES 265

of their Chapter 5, and the Rao-Hamming construction outlined in Section A.8 is given in Chapter 3.

There is an extensive specialized literature on all the topics in this Appendix.

B.10 Further results and exercises

1. For the group $\mathcal{C}_2(a,b,c) = \{1, a, b, ab, c, ac, bc, abc\}$ write out the multiplication table and verify that the group is equally generated by (ab, c, bc). Enumerate all subgroups of $\mathcal{C}_2(a,b,c)$.

2. Write out the multiplication table for $GF(2^2)$.

3. Construct the addition and multiplication table for $GF(3^2)$ taking $x^2 + x + 2$ as the irreducible polynomial. (Verify that it is irreducible.) Verify the power cycle

$$x = x, \quad x^2 = 2x + 1, \quad x^3 = 2x + 2, \quad x^4 = 2,$$
$$x^5 = 2x, \quad x^6 = x + 2, \quad x^7 = x + 1$$

and conversely use the power cycle to derive the multiplication table.

4. Use the addition and multiplication tables for $GF(3^2)$ to write down L_1, L_2 for the 9×9 set. Check that L_2 can be obtained by permuting the rows of L_1.

5. Construct a theory of orthogonal Latin cubes.

6. Count the number of lines and points in finite Euclidean and projective geometries.

APPENDIX C
Computational issues

C.1 Introduction

There is a wide selection of statistical computing packages, and most of these provide the facility for analysis of variance and estimation of treatment contrasts in one form or another. With small data sets it is often straightforward, and very informative, to compute the contrasts of interest by hand. In 2^k factorial designs this is easily done using Yates's algorithm (Exercise 5.1).

The package GENSTAT is particularly well suited to analysis of complex balanced designs arising in agricultural application. SAS is widely used in North America, partly for its capabilities in handling large databases. GLIM is very well suited to empirical model building by the successive addition or deletion of terms, and for analysis of non-normal models of exponential family form.

Because S-PLUS is probably the most flexible and powerful of the packages we give here a very brief overview of the analysis of the more standard designs using S-PLUS, by providing code sufficient for the analysis of the main examples in the text. The reader needing an introduction to S-PLUS or wishing to exploit its full capabilities will need to consult one of the several specialized books on the topic. As with many packaged programs, the output from S-PLUS is typically not in a form suitable for the presentation of conclusions, an important aspect of analysis that we do not discuss.

We assume the reader is familiar with running S-PLUS on the system being used and with the basic structure of S-PLUS, including the use of .Data and CHAPTER directories (the latter introduced in S-PLUS 5.1 for Unix), and the use of *objects* and *methods* for objects. A dataset, a fitted regression model, and a residual plot are all examples of objects. Example of methods for these objects are summary, plot and residuals. Many objects have several specific methods for them as well. The illustrations below use a command line version of S-PLUS such as is often used in a Unix environment.

Most PC based installations of S-PLUS also offer a menu-driven version.

C.2 Overview

C.2.1 Data entry

The typical data from the types of experiments we describe in this book takes a single response or dependent variable at a time, several classification variables such as blocks, treatments, factors and so on, and possibly one or more continuous explanatory variables, such as baseline measurements. The dependent and explanatory variables will typically be entered from a terminal or file, using a version of the `scan` or `read.table` command. It will rarely be the case that the data set will contain fields corresponding to the various classification factors. These can usually be constructed using the `rep` command. All classification or factor variables must be explicitly declared to be so using the `factor` command.

Classification variables for several standard designs can be created by the `fac.design` command. It is usually convenient to collect these classification variables in a `design` object, which essentially contains all the information needed to construct the matrix for the associated linear model.

The collection of explanatory, baseline, and classification variables can be referred to in a variety of ways. The simplest, though in the long run most cumbersome, is to note that variables are automatically saved in the current working directory by the names they are assigned as they are read or created. In this case the data variables relevant to a particular analysis will nearly always be vectors with length equal to the number of responses. Alternatively, when the data file has a spreadsheet format with one row per case and one column per variable, it is often easy to store the dependent and explanatory variables as a matrix. The most flexible and ultimately powerful way to store the data is as a `data.frame`, which is essentially a matrix with rows corresponding to observations and columns corresponding to variables, and a provision for assigning names to the individual columns and rows.

In the first example below we illustrate these three methods of defining and referring to variables: as vectors, as a matrix, and as a data frame. In subsequent examples we always combine the

OVERVIEW

variables in a data frame, usually using a design object for the explanatory variables.

As will be clear from the first example, one disadvantage of a data frame is that individual columns must be accessed by the slightly cumbersome form `data.frame.name$variable.name`. By using the command `attach(data.frame,1)`, the data frame takes the place of the working directory on the search path, and any use of `variable.name` refers to that variable in the attached data frame. More details on the search path for variables is provided by Spector (1994).

C.2.2 Treatment means

The first step in an analysis is usually the construction of a table of treatment means. These can be obtained using the `tapply` command, illustrated in Section C.3 below. To obtain the mean response of y at each of several levels of x use `tapply(y, x, mean)`. In most of our applications x will be a factor variable, but in any case the elements of x are used to define categories for the calculation of the mean. If x is a list then cross-classified means are computed; we use this in Section C.5. In Section C.3 we illustrate the use of `tapply` on a variable, on a matrix, and on a data frame.

A data frame that contains a design object or a number of factor variables has several specialized plotting methods, the most useful of which is `interaction.plot`. Curiously, a summary of means of a design object does not seem to be available, although these means are used by the plotting methods for design objects.

An analysis of variance will normally be used to provide estimated standard errors for the treatment means, using the `aov` command described in the next subsection. If the design is completely balanced, the `model.tables` command can be used on the result of an `aov` command to construct a table of means after an analysis of variance, and this, while in principle not a good idea, will sometimes be more convenient than constructing the table of means before fitting the analysis of variance. For unbalanced or incomplete designs, `model.tables` will give estimated effects, but they are not always properly adjusted for lack of orthogonality.

C.2.3 Analysis of variance

Analysis of variance is carried out using the `aov` command, which is a specialization of the `lm` command used to fit a linear model. The summary and plot methods for `aov` are designed to provide the information most often needed when analysing these kinds of data.

The input to the `aov` command is a response variable and a model formula. S-PLUS has a powerful and flexible modelling language which we will not discuss in any detail. The model formulae for most analyses of variance for balanced designs are relatively straightforward. The model formula takes the form `y ~ model`, where y is the response or dependent variable. Covariates enter `model` by their names only and an overall mean term (denoted `1`) is always assumed to be present unless explicitly deleted from the model formula. If A and B are factors `A + B` represents an additive model with the main effects of A and B, `A:B` represents their interaction, and `A*B` is shorthand for `A + B + A:B`. Thus the linear model

$$E(Y_{js}) = \mu + \beta x_{js} + \tau_j^A + \tau_s^B + \tau_{js}^{AB}$$

can be written

`y~x+A*B`

while

$$E(Y_{js}) = \mu + \beta_j x_{js} + \tau_j^A + \tau_s^B + \tau_{js}^{AB}$$

can be written

`y~x+x:A+A*B.`

There is also a facility for specifying nested effects; for example the model $E(Y_{a;j}) = \mu + \tau_a + \eta_{aj}$ is specified as `y ~ A+B/A`.

Model formulae are discussed in detail by Chambers and Hastie (1992, Chapter 2).

The analysis of variance table is printed by the `summary` function, which takes as its argument the name of the `aov` object. This will show sums of squares corresponding to individual terms in the model. The `summary` command does not show whether or not the sums of squares are adjusted for other terms in the model. In balanced cases the sums of squares are not affected by other terms in the model but in unbalanced cases or in more general models where the effects are not orthogonal, the interpretation of individual sums of squares depends crucially on the other terms in the model.

OVERVIEW

S-PLUS computes the sums of squares much in the manner of stagewise fitting described in Appendix A, and it is also possible to update a fitted model using special notation described in Chambers and Hastie (1992, Chapter 2). The convention is that terms are entered into the model in the order in which they appear on the right hand side of the model statement, so that terms are adjusted for those appearing above it in the **summary** of the **aov** object. For example,

```
unbalanced.aov <- aov(y ~ x1 + x2 + x3)
summary(unbalanced.aov)
```

will fit the models

$$y = \mu + \beta_1 x_1$$
$$y = \mu + \beta_1 x_1 + \beta_2 x_2$$
$$y = \mu + \beta_1 x_1 + \beta_2 x_2 + \beta_3 x_3$$

and in the partitioning of the regression sum of squares the sum of squares attributed to x_1 will be unadjusted, that for x_2 will be adjusted for x_1, and that for x_3 adjusted for x_1 and x_2. Be warned that this is not flagged in the output except by the order of the terms:

```
> summary(unbalanced.aov)
          Df   Sum of Sq   Mean Sq   F Value   Pr(F)
    x1         (unadj.)
    x2         (adj. for x1)
    x3         (adj. for x1, x2)
residuals
```

C.2.4 Contrasts and partitioning sums of squares

As outlined in Section 3.5, it is often of interest to partition the sums of squares due to treatments using linear contrasts. In S-PLUS each factor variable has an associated set of linear contrasts, which are used as parameterization constraints in the fitting of the model specified in the **aov** command. These linear contrasts determine the estimated values of the unknown parameters. They can also be used to partition the associated sum of squares in the analysis of variance table using the **split** option to **summary(aov)**.

This dual use of contrasts for factor variables is very powerful, although somewhat confusing. We will first indicate the use of contrasts in estimation, before using them to partition the sums of squares.

The default contrasts for an unordered factor, which is created by `factor(x)`, are *Helmert contrasts*, which compare the second level with the first, the third level with the average of the first two, and so on. The default contrasts for an ordered factor are those determined by the appropriate orthogonal polynomials. The contrasts used in fitting can be changed before an analysis of variance is constructed, using the `options` command. The summation constraint for an unordered factor $\sum \tau_j = 0$ is imposed by specifying `contr.sum`, and the constraint $\tau_1 = 0$ is imposed by specifying `contr.treatment`:

```
> options(contrasts=c("contr.sum", "contr.poly"))
> options(contrasts=c("contr.treatment", "contr.poly"))
```

It is possible to specify a different set of contrasts for ordered factors from polynomial contrasts, but this will rarely be needed. In Section C.3.3 below we estimate the treatment parameters under each of the three constraints: Helmert, summation and $\tau_1 = 0$. If individual estimates of the τ_j are to be used for any purpose, and this should be avoided as far as feasible, it is essential to note the constraints under which these estimates were obtained.

The contrasts used in fitting the model can also be used to partition the sums of squares. The summation contrasts will rarely be of interest in this context, but the orthogonal polynomial contrasts will be useful for quantitative factors. Prespecified contrasts may also be specified, using the function `contrasts` or `C`. Use of the contrast matrix `C` is outlined in detail by Venables and Ripley (1999, Chapter 6.2).

C.2.5 Plotting

There are some associated plotting methods that are often useful. The function `interaction.plot` plots the mean response by levels of two cross-classified factors, and is illustrated in Section C.5 below. An optional argument allows some other function of the response, such as the median or standard error, to be plotted instead.

The function `qqnorm`, when applied to an analysis of variance object created by the `aov` command, constructs a half-normal plot of the estimated effects (see Section 5.5). Two optional arguments are very useful: `qqnorm(aov.example, label=6)` will label the six

RANDOMIZED BLOCK EXPERIMENT FROM CHAPTER 3   273

largest effects on the plot, and `qqnorm(aov.example, full=T)` will construct a full normal plot of the estimated effects.

C.2.6 Specialized commands for standard designs

There are a number of commands for constructing designs, including `fac.design, oa.design`, and `expand.grid`. Fractional factorials can be constructed with an optional argument to `fac.design`. Details on the use of these functions are given by Chambers and Hastie (1992, Chapter 5.2); see also Venables and Ripley (1999, Chapter 6.7).

C.2.7 Missing values

Missing values are generally assigned the special value `NA`. S-PLUS functions differ in their handling of missing values. Many of the plotting functions, for example, will plot missing values as zeroes; the documentation for, for example, `interaction.plot` includes under the description of the response variable the information "Missing values (NA) are allowed". On the other hand, `aov` handles missing values in the same way `lm` does, through the optional argument `na.action`. The default value for `na.action` is `na.fail`, which halts further computation. Two alternatives are `na.omit`, which will omit any rows of the data frame that have missing values, and `na.include`, which will treat `NA` as a valid factor level among all factor variables; see Spector (1994, Ch. 10).

In some design and analysis textbooks there are formulae for computing (by hand) treatment contrasts, standard errors, and analysis of variance tables in the presence of a small number of missing responses in randomized block designs; Cochran and Cox (1958) provide details for a number of other more complex designs. In general, procedures for arbitrarily unbalanced data may have to be used.

C.3 Randomized block experiment from Chapter 3

C.3.1 Data entry

This is the randomized block experiment taken from Cochran and Cox (1958), to compare five quantities of potash fertiliser on the strength of cotton fiber. The data and analysis of variance are given in Tables 3.1 and 3.2. The dependent variable is strength, and there are two classification variables, treatment (amount of potash), and block. The simplest way to enter the data is within S-PLUS:

```
> potash.strength<-scan()
1: 762 814 776 717 746 800 815 773 757 768 793 787 774 780 721
16:
> potash.strength<-potash.strength/100
> potash.tmt<-factor(rep(1:5,3))
> potash.blk<-factor(rep(1:3,rep(5,3)))
> potash.tmt
 [1] 1 2 3 4 5 1 2 3 4 5 1 2 3 4 5
> potash.blk
 [1] 1 1 1 1 1 2 2 2 2 2 3 3 3 3 3
> is.factor(potash.tmt)
[1] T
```

We could also construct a 15×3 matrix to hold the response variable and the explanatory variables, although the columns of this matrix are all considered numeric, even if the variable entered is a factor.

```
> potash.matrix<-matrix(c(potash.strength, potash.tmt,
+ potash.blk),15,3)
> potash.matrix
      [,1] [,2] [,3]
 [1,] 7.62   1    1
 [2,] 8.14   2    1
 [3,] 7.76   3    1
 [4,] 7.17   4    1
         .
         .
         .
>is.factor(potash.tmt)
[1] T
>is.factor(potash.matrix[,2])
[1] F
```

Finally we can construct the factor levels using `fac.design`, store them in the design object `potash.design`, and combine this with the dependent variable in a data frame `potash.df`. In the illustration below we add 'names' for the factor levels, an option that is available (but not required) in the `fac.design` command.

```
> fnames<-list(tmt=c("36", "54", "72", "108", "144"),
+ blk=c("I","II","III"))
> potash.design<-fac.design(c(5,3),fnames)
> potash.design
   tmt blk
1   36   I
2   54   I
3   72   I
4  108   I
```

```
                    .
                    .
                    .
> strength<-potash.strength  # this is simply to use a shorter
                             # name in what follows
> rm(strength, fnames, potash.design) # remove un-needed objects
> potash.df<-data.frame(strength,potash.design)
> potash.df
 potash.df
    strength tmt blk
 1      7.62  36  I
 2      8.14  54  I
 3      7.76  72  I
 4      7.17 108  I
                    .
                    .
                    .
```

C.3.2 Table of treatment and block means

The simplest way to compute the treatment means is using the `tapply` command. When used with an optional factor argument as `tapply(y,factor,mean)` the calculation of the mean is stratified by the level of the factor. This can be used on any of the data structures outlined in the previous subsection:

```
> tapply(potash.strength,potash.tmt,mean)
     1      2      3      4      5
  7.85 8.0533 7.7433 7.5133   7.45

> tapply(potash.matrix[,1],potash.matrix[,2],mean)
     1      2      3      4      5
  7.85 8.0533 7.7433 7.5133   7.45

> tapply(potash.df$strength, potash.df$tmt, mean)
    36     54     72    108    144
  7.85 8.0533 7.7433 7.5133   7.45
```

As is apparent above, the `tapply` command is not terribly convenient when used on a data matrix or a data frame. There are special plotting methods for data frames with factors that allow easy plotting of the treatment means, but curiously there does not seem to be a ready way to print the treatment means without first constructing an analysis of variance.

C.3.3 Analysis of variance

We first form a two way analysis of variance using `aov`. Note that the `summary` method for the analysis of variance object gives more useful output than printing the object itself.

In this example we illustrate the estimates $\hat{\tau}_j$ in the model $y_{js} = \mu + \tau_j + \beta_s + \epsilon_{js}$ under the default constraint specified by the Helmert contrasts, under the summation constraint $\sum \tau_j = 0$, and under the constraint often used in generalized linear models $\tau_1 = 0$. If individual estimates of the τ_j are to be used for any purpose, it is essential to note the constraints under which these estimates were obtained. The analysis of variance table and estimated residual sum of squares are of course invariant to the choice of parametrization constraint.

```
> potash.aov<-aov(strength~tmt+blk,potash.df)
> potash.aov
Call:
   aov(formula = strength ~ tmt + blk, data = potash.df)

Terms:
                     tmt     blk Residuals
 Sum of Squares  0.73244 0.09712   0.34948
 Deg. of Freedom       4       2         8

Residual standard error: 0.20901
Estimated effects are balanced
> summary(potash.aov)
          Df Sum of Sq Mean Sq F Value   Pr(F)
      tmt  4   0.73244 0.18311  4.1916 0.04037
      blk  2   0.09712 0.04856  1.1116 0.37499
Residuals  8   0.34948 0.04369

> coef(potash.aov)
 (Intercept)      tmt1       tmt2      tmt3    tmt4   blk1   blk2
       7.722   0.10167  -0.069444 -0.092222  -0.068  0.098 -0.006

> options(contrasts=c("contr.sum","contr.poly"))
> potash.aov<-aov(strength~tmt+blk,potash.df)
> coef(potash.aov)
 (Intercept)   tmt1    tmt2      tmt3     tmt4    blk1  blk2
       7.722  0.128 0.33133  0.021333 -0.20867  -0.092 0.104

> options(contrasts=c("contr.treatment","contr.poly"))
> potash.aov<-aov(strength~tmt+blk,potash.df)
> coef(potash.aov)
 (Intercept)    tmt54     tmt72    tmt108 tmt144 blkII blkIII
       7.758  0.20333  -0.10667  -0.33667   -0.4 0.196   0.08
```

The estimated treatment effects under the summation constraint can also be obtained using `model.tables` or `dummy.coef`, so it

RANDOMIZED BLOCK EXPERIMENT FROM CHAPTER 3

is not necessary to change the default fitting constraint with the `options` command, although it is probably advisable. Below we illustrate this, assuming that the default (Helmert) contrasts were used in the `aov` command. We also illustrate how `model.tables` can be used to obtain treatment means and their standard errors.

```
> options("contrasts")
$contrasts:
[1] "contr.helmert"  "contr.poly"

> dummy.coef(potash.aov)
$"(Intercept)":
[1] 7.722

$tmt:
    36       54        72       108      144
 0.128  0.33133  0.021333 -0.20867  -0.272

$blk:
     I     II     III
-0.092  0.104  -0.012

> model.tables(potash.aov)

Tables of effects

  tmt
        36        54       72      108       144
   0.12800  0.33133  0.02133 -0.20867  -0.27200

  blk
       I      II     III
  -0.092   0.104  -0.012
Warning messages:
  Model was refitted to allow projection in: model.tables(potash.aov)

> model.tables(potash.aov,type="means",se=T)

Tables of means
Grand mean

 7.722

  tmt
       36      54      72     108     144
   7.8500  8.0533  7.7433  7.5133  7.4500

  blk
       I      II     III
   7.630   7.826   7.710

Standard errors for differences of means
              tmt        blk
```

```
           0.17066 0.13219
replic. 3.00000 5.00000
```

C.3.4 Partitioning sums of squares

For the potash experiment, the treatment was a quantitative factor, and in Section 3.5.5 we discussed partitioning the treatment sums of squares using the linear and quadratic polynomial contrasts for a factor with five levels using $(-2, -1, 0, 1, 2)$ and $(2, -1, -2, -1, 2)$. Since orthogonal polynomials are the default for an ordered factor, the simplest way to partition the sums of squares in S-PLUS is to define tmt as an ordered factor.

```
> otmt<-ordered(potash.df$tmt)
> is.ordered(otmt)
[1] T
> is.factor(otmt)
[1] T
> contrasts(otmt)
             .L          .Q          .C         ^ 4
 36 -6.3246e-01   0.53452 -3.1623e-01    0.11952
 54 -3.1623e-01  -0.26726  6.3246e-01   -0.47809
 72 -6.9389e-18  -0.53452  4.9960e-16    0.71714
108  3.1623e-01  -0.26726 -6.3246e-01   -0.47809
144  6.3246e-01   0.53452  3.1623e-01    0.11952
> potash.df<-data.frame(potash.df,otmt)
> rm(otmt)
> potash.aov<-aov(strength~otmt+blk,potash.df)
> summary(potash.aov)
           Df Sum of Sq  Mean Sq  F Value    Pr(F)
      otmt  4   0.73244  0.18311   4.1916  0.04037
       blk  2   0.09712  0.04856   1.1116  0.37499
 Residuals  8   0.34948  0.04369
> summary(potash.aov,split=list(otmt=list(L=1,Q=2)))
           Df Sum of Sq  Mean Sq  F Value    Pr(F)
      otmt  4   0.73244  0.18311    4.192  0.04037
   otmt: L  1   0.53868  0.53868   12.331  0.00794
   otmt: Q  1   0.04404  0.04404    1.008  0.34476
       blk  2   0.09712  0.04856    1.112  0.37499
 Residuals  8   0.34948  0.04369
> summary(potash.aov,split=list(otmt=list(L=1,Q=2,C=3,QQ=4)))
           Df Sum of Sq  Mean Sq  F Value    Pr(F)
      otmt  4   0.73244  0.18311    4.192  0.04037
   otmt: L  1   0.53868  0.53868   12.331  0.00794
   otmt: Q  1   0.04404  0.04404    1.008  0.34476
   otmt: C  1   0.13872  0.13872    3.175  0.11261
  otmt: QQ  1   0.01100  0.01100    0.252  0.62930
       blk  2   0.09712  0.04856    1.112  0.37499
 Residuals  8   0.34948  0.04369
```

It is possible to specify just one contrast of interest, and a set of

RANDOMIZED BLOCK EXPERIMENT FROM CHAPTER 3

contrasts orthogonal to the first will be constructed automatically. This set will not necessarily correspond to orthogonal polynomials however.

```
> contrasts(tmt)<-c(-2,-1,0,1,2)
> contrasts(tmt)          #these contrasts are orthogonal
                   #but not the usual polynomial contrasts
      [,1]      [,2]      [,3]     [,4]
36    -2  -0.41491  -0.3626  -0.3104
54    -1   0.06722   0.3996   0.7320
72     0   0.83771  -0.2013  -0.2403
108    1  -0.21744   0.6543  -0.4739
144    2  -0.27258  -0.4900   0.2925

> potash.aov<-aov(strength~tmt+blk,potash.df)
> summary(potash.aov,split=list(tmt=list(1)))
             Df Sum of Sq Mean Sq F Value   Pr(F)
       tmt    4    0.7324  0.1831    4.19  0.0404
tmt: Ctst 1   1    0.5387  0.5387   12.33  0.0079
       blk    2    0.0971  0.0486    1.11  0.3750
 Residuals    8    0.3495  0.0437
```

Finally, in this example recall that the treatment levels are not in fact equally spaced, so that the exact linear contrast is as given in Section 3.5: $(-2, -1.23, -0.46, 1.08, 2.6)$. This can be specified using `contrasts`, as illustrated here.

```
> contrasts(potash.tmt)<-c(-2,-1.23,-0.46,1.08,2.6)
> contrasts(potash.tmt)
      [,1]      [,2]      [,3]      [,4]
1    -2.00  -0.44375  -0.4103  -0.3773
2    -1.23  -0.09398   0.3332   0.7548
3    -0.46   0.86128  -0.1438  -0.1488
4     1.08  -0.15416   0.6917  -0.4605
5     2.60  -0.16939  -0.4707   0.2318
> potash.aov<-aov(potash.strength~potash.tmt+potash.blk)
> summary(potash.aov,split=list(potash.tmt=list(1,2,3,4)))
                      Df Sum of Sq Mean Sq F Value   Pr(F)
       potash.tmt      4    0.7324  0.1831    4.19  0.0404
potash.tmt: Ctst 1     1    0.5668  0.5668   12.97  0.0070
potash.tmt: Ctst 2     1    0.0002  0.0002    0.01  0.9444
potash.tmt: Ctst 3     1    0.0045  0.0045    0.10  0.7577
potash.tmt: Ctst 4     1    0.1610  0.1610    3.69  0.0912
       potash.blk      2    0.0971  0.0486    1.11  0.3750
        Residuals      8    0.3495  0.0437
```

C.4 Analysis of block designs in Chapter 4

C.4.1 Balanced incomplete block design

The first example in Section 4.2.6 is a balanced incomplete block design with two treatments per block in each of 15 blocks. The data are entered as follows:

```
> weight<-scan()
1: 251 215 249 223 254 226 258 215 265 241
11: 211 190 228 211 215 170 232 253 215 223
21: 234 215 230 249 220 218 226 243 228 256
31:
> weight<-weight/100
> blk<-factor(rep(1:15,rep(2,15)))
> blk
 [1] 1 1 2 2 3 3 4 4 ...
> tmt <- 0
> for (i in 1:5) for (j in (i+1):6) tmt <- c(tmt,i,j)
> tmt <- tmt[-1]
> tmt <- factor(tmt)
> tmt
 [1] 1 2 1 3 1 4 1 5 1 6 2 3 2 4 2 5 2 6 3 4 3 5 3 6 4 5 4 6 5 6
> fnames<-list(tmt=c("C","His-","Arg-","Thr-","Val-","Lys-"),
+ blk=c(1:15))
> chick.design<-design(tmt,blk,factor.names=fnames)
> chick.design
    tmt blk
1    C   1
2 His-   1
3    C   2
4 Arg-   2
     .
     .
     .
> chick.df<-data.frame(weight,chick.design)
> rm(chick.design, fnames, blk)
> chick.df
  weight  tmt blk
1   2.51    C   1
2   2.15 His-   1
3   2.49    C   2
4   2.23 Arg-   2
5   2.54    C   3
6   2.26 Thr-   3
     .
     .
     .
```

We now compute treatment means, both adjusted and unadjusted, and the analysis of variance table for their comparison. This is our first example of an unbalanced design, in which for example the sums of squares for treatments ignoring blocks is different from

ANALYSIS OF BLOCK DESIGNS IN CHAPTER 4   281

the sums of squares adjusted for blocks. The convention in S-PLUS is that terms are added to the model in the order they are listed in the model statement. Thus to construct the intrablock analysis of variance, in which treatments are adjusted for blocks, we use the model statement y ~ block + treatment.

We used tapply to obtain the unadjusted treatment means, and obtained the adjusted means by adding $\hat{\tau}_j$ to the overall mean $\bar{Y}_{..}$. The $\hat{\tau}_j$ were obtained under the summation constraint. It is possible to derive both Q_j and the adjusted treatment means using model.tables, although this returns an incorrect estimate of the standard error and is not recommended. The least squares estimates of τ_j under the summation constraint are also returned by dummy.coef, even if the summation constraint option was not specified in fitting the model.

```
> tapply(weight,tmt,mean)
    1     2     3     4     5     6
2.554 2.202 2.184 2.212 2.092 2.484
> options(contrasts=c("contr.sum","contr.poly"))
> chick.aov<-aov(weight~blk+tmt,chick.df)
> summary(chick.aov)
          Df  Sum of Sq  Mean Sq  F Value     Pr(F)
      blk 14    0.75288  0.053777  8.173  0.0010245
      tmt  5    0.44620  0.089240 13.562  0.0003470
Residuals 10    0.06580  0.006580

> coef(chick.aov)
 (Intercept)     blk1     blk2    blk3     blk4     blk5     blk6
       2.288  -0.1105  -0.013  0.0245 0.060333 0.060333 -0.25883

      blk7      blk8    blk9     blk10  blk11 blk12   blk13  blk14
 -0.071333 -0.2705  0.0645 -0.0088333  0.117 0.102  0.0595 0.0495

     tmt1      tmt2      tmt3      tmt4     tmt5
  0.26167  0.043333 -0.091667 -0.086667 -0.22833

> dummy.coef(chick.aov)
$"(Intercept)":
[1] 2.288
 ...
$tmt:
       C      His-      Arg-      Thr-      Val-     Lys-
 0.26167  0.043333 -0.091667 -0.086667 -0.22833  0.10167

> tauhat<-.Last.value$tmt
> tauhat+mean(weight)
     C    His-    Arg-    Thr-    Val-    Lys-
2.5497  2.3313  2.1963  2.2013  2.0597  2.3897

> model.tables(chick.aov,type="adj.means")
```

```
Tables of adjusted means
Grand mean

   2.28800
se 0.01481

...

  tmt
         C    His-   Arg-   Thr-   Val-   Lys-
    2.5497 2.3313 2.1963 2.2013 2.0597 2.3897
se  0.0452 0.0452 0.0452 0.0452 0.0452 0.0452
```

We will now compute the interblock analysis of variance using regression on the block totals. The most straightforward approach is to compute the estimates directly from equations (4.32) and (4.33); the estimated variance is obtained from the analysis of variance table with blocks adjusted for treatments. To obtain this analysis of variance table we specify treatment first in the right hand side of the model statement that is the argument of the `aov` command.

```
> N <- matrix(0, nrow=6, ncol=15)
> ind <- 0
> for (i in 1:5) for (j in (i+1):6) ind <- c(ind,i,j)
> ind<- ind[-1]
> ind <- matrix(ind, ncol=2,byrow=T)
> for (i in 1:15) N[ind[i,1],i] <- N[ind[i,2],i] <-1
> B<-tapply(weight,blk,sum)
> B
   1    2    3    4    5    6    7    8    9   10   11   12
4.66 4.72 4.8  4.73 5.06 4.01 4.39 3.85 4.85 4.38 4.49 4.79

  13   14   15
4.38 4.69 4.84

> tau<-(N%*%B-5*2*mean(weight))/4
> tau<-as.vector(tau)
> tau
[1]  0.2725 -0.2800 -0.1225 -0.0600 -0.1475  0.3375
>
> summary(aov(weight~tmt+blk,chick.df))
          Df Sum of Sq Mean Sq F Value    Pr(F)
      tmt  5   0.85788 0.17158  26.075 0.000020
      blk 14   0.34120 0.02437   3.704 0.021648
Residuals 10   0.06580 0.00658
> sigmasq<-0.00658
> sigmaBsq<-((0.34120/14-0.00658)*14)/(6*4)
> sigmaBsq
[1] 0.010378
> vartau1<-sigmasq*2*5/(6*6)
> vartau2<-(2*5*(sigmasq+2*sigmaBsq))/(6*4)
> (1/vartau1)+(1/vartau2)
```

```
[1] 634.91
> (1/vartau1)/.Last.value
[1] 0.86172
> dummy.coef(chick.aov)$tmt
        C     His-       Arg-      Thr-       Val-      Lys-
  0.26167 0.043333 -0.091667 -0.086667 -0.22833 0.10167
> tauhat<-.Last.value
> taustar<-.86172*tauhat+(1-.86172)*tau
> taustar
        C      His-       Arg-       Thr-      Val-      Lys-
  0.26316 -0.0013772 -0.09593 -0.082979 -0.21716 0.13428
> sqrt(1/( (1/vartau1)+(1/vartau2)))
[1] 0.039687
> setaustar<-.Last.value
> sqrt(2)*setaustar
[1] 0.056125
```

C.4.2 Unbalanced incomplete block experiment

The second example from Section 4.2.6 has all treatment effects highly aliased with blocks. The data is given in Table 4.13 and the analysis summarized in Tables 4.14 and 4.15. The within block analysis is computed using the aov command, with blocks (days) entered into the model before treatments. The adjusted treatment means are computed by adding $\bar{Y}_{..}$ to the estimated coefficients. We also indicate the computation of the least squares estimates under the summation constraint using the matrix formulae of Section 4.2. The contrasts between pairs of treatment means do not have equal precision; the estimated standard error is computed for each mean using $\mathrm{var}(\bar{Y}_{j.}) = \sigma^2/r_j$, although for comparing pairs of means it may be more useful to use the result that $\mathrm{cov}(\hat{\tau}) = C^-$.

```
> day<-rep(1:7,rep(4,7))
> tmt<-scan()
1: 1 8 9 9 9 5 4 9 2 3 8 5 ...
29:
> expansion<-scan()
1: 150 148 130 117 122 141 112 ...
29:
> day<-factor(day)
> tmt<-factor(tmt)
> expansion<-expansion/10
> dough.design<-design(tmt,day)
> dough.df<-data.frame(expansion,dough.design)
> dough.df
  expansion tmt day
1      15.0   1   1
2      14.8   8   1
3      13.0   9   1
```

```
4       11.7    9   1
5       12.2    9   2
6       14.1    5   2
                .
                .
                .
```

```
> tapply(expansion,day,mean)
     1      2      3      4      5      6     7
13.625 12.275 14.525 13.475 11.475 15.15 11.55
> tapply(expansion,tmt,mean)
   1     2     3  4     5    6     7    8        9    10 11
14.8 15.45 10.45 12 14.85 18.2 13.15 15.3 11.46667 11.2 13

  12   13   14   15
12.7 11.7 11.4 11.1

> dough.aov<-aov(expansion~day+tmt,dough.df)
> summary(dough.aov)
          Df Sum of Sq Mean Sq F Value   Pr(F)
      day  6     49.41   8.235  11.188 0.00275
      tmt 14     96.22   6.873   9.337 0.00315
Residuals  7      5.15   0.736

> dummy.coef(.Last.value)$tmt
      1      2       3       4      5      6       7      8
 1.3706 3.5372 -2.3156 -1.0711 2.1622 3.9178 0.85389 2.2539

        9      10      11      12       13      14      15
 -0.51556 -3.4822 0.58444 -1.9822 -0.71556 -3.2822 -1.3156

> replications(dough.design)
$tmt:
 1 2 3 4 5 6 7 8 9 10 11 12 13 14 15
 2 2 1 2 2 1 1 6 1  1  2  2  1  2  2

$day:
[1] 4
> R<-matrix(0,nrow=15,ncol=15)
> diag(R)<-replications(dough.design)$tmt
> K<-matrix(0,nrow=7,ncol=7)
> diag(K)<-rep(4,7)
> N<-matrix(0,nrow=15,ncol=7)
> N[,1]<-c(1,0,0,0,0,0,0,1,2,0,0,0,0,0,0)
> N[,2]<-c(0,0,0,1,1,0,0,0,2,0,0,0,0,0,0)
> N[,3]<-c(0,1,1,0,1,0,0,1,0,0,0,0,0,0,0)
> N[,4]<-c(0,0,0,0,0,1,0,0,0,1,0,1,0,1,0)
> N[,5]<-c(0,0,1,0,0,0,0,0,0,0,1,0,1,0,1)
> N[,6]<-c(1,0,0,1,0,1,1,0,0,0,0,0,0,0,0)
> N[,7]<-c(0,1,0,0,0,0,1,0,2,0,0,0,0,0,0)
> Q<-S-N%*%solve(K)%*%B
> C<-R-N%*%solve(K)%*%t(N)
> Q%*%ginverse(C)
         [,1]    [,2]    [,3]    [,4]    [,5]    [,6]    [,7]    [,8]
```

EXAMPLES FROM CHAPTER 5   285

```
[1,]  1.3706  3.5372 -2.3156 -1.0711  2.1622  3.9178  0.85389  2.2539

           [,9]    [,10]   [,11]    [,12]    [,13]    [,14]    [,15]
[1,]   -0.51556 -3.4822 0.58444 -1.9822 -0.71556 -3.2822 -1.3156
> tauhat<-.Last.value
> as.vector(tauhat+mean(expansion))
 [1] 14.5241 16.6908 10.8380 12.0825 15.3158 17.0713 14.0075
 [8] 15.4075 12.6380  9.6713 13.7380 11.1713 12.4380  9.8713
[15] 11.8380
> se<-0.7361/sqrt(diag(R))
> se
 [1] 0.52050 0.52050 0.52050 0.52050 0.52050 0.52050 0.52050
 [8] 0.52050 0.30051 0.73610 0.73610 0.73610 0.73610 0.73610
[15] 0.73610
> setauhat<-sqrt(diag(ginverse(C)))
> setauhat
 [1] 0.92376 0.92376 1.04243 0.92376 0.92376 1.04243 0.92376
 [8] 0.92376 0.76594 1.59792 1.59792 1.59792 1.59792 1.59792
[15] 1.59792
```

C.5 Examples from Chapter 5

C.5.1 Factorial experiment, Section 5.2

The treatments in this experiment form a complete $3 \times 2 \times 2$ factorial. The data are given in Table 5.1 and the analysis summarized in Tables 5.2 and 5.4. The code below illustrates the use of `fac.design` to construct the levels of the factors. For this purpose we treat house as a factor, although in line with the discussion of Section 5.1 it is not an aspect of treatment. These factors are then used to stratify the response in the `tapply` command, producing tables of marginal means. Figure 5.1 was obtained using `interaction.plot`, after constructing a four-level factor indexing the four combinations of type of protein crossed with level of fish solubles.

```
> weight<-scan()
1: 6559 6292 7075 6779 6564 6622 7528 6856 6738 6444 7333 6361
13: 7094 7053 8005 7657 6943 6249 7359 7292 6748 6422 6764 6560
25:

> exk.design<-fac.design(c(2,2,3,2),factor.names=
+ list(House=c("I","II"), Lev.f=c("0","1"),
+ Lev.pro=c("0","1","2"),Type=c("gnut","soy")))
> exk.design
  House Lev.f Lev.pro Type
1     I     0       0 gnut
2    II     0       0 gnut
3     I     1       0 gnut
```

```
4    II    1    0 gnut
5    I     0    1 gnut
...
```

```
> exk.df<-data.frame(weight,exk.design)
> rm(exk.design)
> tapply(weight,list(exk.df$Lev.pro,exk.df$Type),mean)
    gnut    soy
0 6676.2 7452.2
1 6892.5 6960.8
2 6719.0 6623.5

> tapply(weight,list(exk.df$Lev.f,exk.df$Type),mean)
    gnut    soy
0 6536.5 6751.5
1 6988.7 7272.8

> tapply(weight,list(exk.df$Lev.f,exk.df$Lev.pro),mean)
       0      1      2
0 6749.5 6594.5 6588.0
1 7379.0 7258.7 6754.5

> tapply(weight, list(exk.df$Lev.pro,exk.df$Lev.f,exk.df$Type),
+ mean)
, , gnut
       0    1
0 6425.5 6927
1 6593.0 7192
2 6591.0 6847

, , soy
       0      1
0 7073.5 7831.0
1 6596.0 7325.5
2 6585.0 6662.0

> Type.Lev.f<-factor(c(1,1,2,2,1,1,2,2,1,1,2,2,
+ 3,3,4,4,3,3,4,4,3,3,4,4))
> postscript(file="Fig5.1.ps",horizontal=F)
> interaction.plot(exk.df$Lev.pro,Type.Lev.f,weight,
+ xlab="Level of protein")
> dev.off()
```

Table 5.3 shows the analysis of variance, using interactions with houses as the estimate of error variance. As usual, the summary table for the analysis of variance includes calculation of F statistics and associated p-values, whether or not these make sense in light of the design. For example, the F statistic for the main effect of houses does not have a justification under the randomization, which was limited to the assignment of chicks to treatments. Indi-

EXAMPLES FROM CHAPTER 5

vidual assessment of main effects and interactions via F-tests is also usually not relevant; the main interest is in comparing treatment means. As the design is fully balanced, `model.tables` provides a set of cross-classified means, as well as the standard errors for their comparison. The linear and quadratic contrasts for the three-level factor level of protein are obtained first by defining protein as an ordered factor, and then by using the `split` option to the analysis of variance summary.

```
> exk.aov<-aov(weight~Lev.f*Lev.pro*Type+House,exk.df)
> summary(exk.aov)
                   Df Sum of Sq Mean Sq F Value   Pr(F)
            Lev.f   1   1421553 1421553  31.741 0.00015
          Lev.pro   2    636283  318141   7.104 0.01045
             Type   1    373751  373751   8.345 0.01474
            House   1    708297  708297  15.815 0.00217
    Lev.f:Lev.pro   2    308888  154444   3.449 0.06876
       Lev.f:Type   1      7176    7176   0.160 0.69661
     Lev.pro:Type   2    858158  429079   9.581 0.00390
Lev.f:Lev.pro:Type  2     50128   25064   0.560 0.58686
        Residuals  11    492640   44785

> model.tables(exk.aov,type="mean",se=T)

Tables of means
Grand mean

 6887.4

 Lev.f
    0      1
 6644 7130.8

 ...

Standard errors for differences of means
         Lev.f Lev.pro   Type  House Lev.f:Lev.pro Lev.f:Type
        86.396  105.81 86.396 86.396        149.64     122.18
replic. 12.000    8.00 12.000 12.000          4.00       6.00

        Lev.pro:Type Lev.f:Lev.pro:Type
              149.64             211.63
replic.         4.00               2.00
Warning messages:
  Model was refit to allow projection in: model.tables(exk.aov,
       type = "means", se = T)

> options(contrasts=c("contr.poly","contr.poly"))
> exk.aov2<-aov(weight~Lev.f*Lev.pro*Type + House, data=exk.df)
> summary(exk.aov2,split=list(Lev.pro=list(1,2)))

                                 Df Sum of Sq Mean Sq  F Value   Pr(F)
```

Lev.f	1	1421553	1421553	31.74	0.0001
Lev.pro	2	636283	318141	7.10	0.0104
Lev.pro: Ctst 1	1	617796	617796	13.79	0.0034
Lev.pro: Ctst 2	1	18487	18487	0.41	0.5337
Type	1	373751	373751	8.34	0.0147
House	1	708297	708297	15.81	0.0022
Lev.f:Lev.pro	2	308888	154444	3.45	0.0689
Lev.f:Lev.pro: Ctst 1	1	214369	214369	4.79	0.0512
Lev.f:Lev.pro: Ctst 2	1	94519	94519	2.11	0.1742
Lev.f:Type	1	7176	7176	0.16	0.6966
Lev.pro:Type	2	858158	429079	9.58	0.0039
Lev.pro:Type: Ctst 1	1	759512	759512	16.96	0.0017
Lev.pro:Type: Ctst 2	1	98645	98645	2.20	0.1658
Lev.f:Lev.pro:Type	2	50128	25064	0.56	0.5869
Lev.f:Lev.pro:Type: Ctst 1	1	47306	47306	1.06	0.3261
Lev.f:Lev.pro:Type: Ctst 2	1	2821	2821	0.06	0.8064
Residuals	11	492640	44785		

C.5.2 2^{4-1} fractional factorial; Section 5.7

The data for the nutrition trial of Blot et al. (1993) is given in Table 5.9. Below we illustrate the analysis of the log of the death rate from cancer, and the numbers of cancer deaths. The second analysis is a reasonable approximation to the first; as the numbers at risk are nearly equal across treatment groups. Both these analyses ignore the blocking information on sex, age and commune. Blot et al. (1993) report the results in terms of the relative risk, adjusting for the blocking factors; the conclusions are broadly similar. The fraction option to fac.design defaults to the highest order interaction for defining the fraction. In the model formula the shorthand .^2 denotes all main effects and two-factor interactions. We illustrate the use of qqnorm for constructing a half-normal plot of the estimated effects from an aov object. The command qqnorm.aov is identical to qqnorm. The command qqnorm(aov.object, full=T) will produce a full-normal plot of the estimated effects, and effects other than the grand mean can be omitted from the plot with the option omit=. Here we omitted the plotting of the aliased effects, otherwise they are plotted as 0.

```
> lohi<-c("0","1")
> cancer.design<- fac.design(levels=c(2,2,2,2),
+ factor=list(A=lohi,B=lohi,C=lohi,D=lohi),fraction=1/2)
> death.c<-scan()
1: 107 94 121 101 81 103 90 95
9:
> years<-scan()
1: 18626 18736 18701 18686 18745 18729 18758 18792
```

EXAMPLES FROM CHAPTER 5   289

```
9:
> log.rates<-log(death.c/years)

# Below we analyse number of deaths from cancer and
# the log death rate; the latter is discussed in Section 5.7.

> logcancer.df<-data.frame(log.rates,cancer.design)

> cancer.df<-data.frame(death.c,cancer.design)
> rm(lohi,death.c,log.rates)

> logcancer.aov<-aov(log.rates~.^2,logcancer.df)
> model.tables(logcancer.aov,type="feffects")

Table of factorial effects
          A         B        C        D       A:B       A:C
  -0.036108 -0.005475 0.053475 -0.13972 -0.043246 0.15217

          A:D
  -0.058331
> cancer.aov<-aov(death.c~.^2,cancer.df)
> model.tables(cancer.aov,type="feffects")

Table of factorial effects
    A    B   C    D A:B A:C A:D
 -2.5 -1.5 5.5 -13.5  -5  15  -6

> postscript(file="FigC.1.ps",horizontal=F)
> qqnorm(logcancer.aov,omit=c(8,9,10),label=7)
> dev.off()

> mean(1/death.c)
[1] 0.0102
```

C.5.3 Exercise 5.6: flour milling

This example is adapted from Tuck, Lewis and Cottrell (1993); that article provides a detailed case study of the use of response surface methods in a quality improvement study in the flour milling industry. A subset of the full data from the article's experiment I is given in Table 5.8. There are six factors of interest, all quantitative, labelled XA through XF and coded -1 and 1. (The variable name "F" is reserved in S-PLUS for "False".) The experiment forms a one-quarter fraction of a 2^6 factorial. The complete data included a further 13 runs taken at coded values for the factors arranged in what is called in response surface methodology a central composite design. Below we construct the fractional factorial by specifying the defining relations as an optional argument to `fac.design`. As the S-PLUS default is to vary the first factor most quickly, which is the

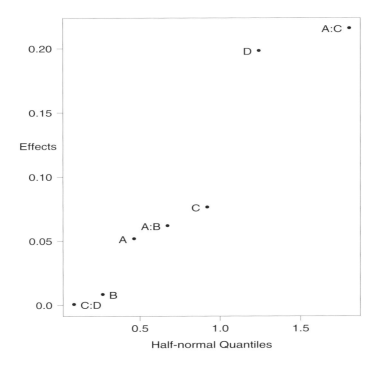

Figure C.1 *Half normal plots of estimated effects: cancer mortality in Linxiang nutrition trial.*

opposite of the design given in Table 5.8, we name the factors in reverse order.

```
> flour.y <- scan()
1: 519 446 337 415 503 468 343 418 ...

61: 551 500 373 462
65:
> flour.tmt <- rep(1:16,rep(4,16))
> flour.tmt
 [1]  1  1  1  1  2  2  2  2  3  3  3  3 ...

> flour.tmt <- factor(flour.tmt)
> flour.day <- rep(1:4,16)
> tapply(flour.y,flour.tmt,mean)
      1      2      3      4      5      6      7      8      9
 429.25    433 454.25 456.75 446.75 447.75  455.5 448.25 458.75
     10     11     12     13     14     15     16
  449.5 463.75    386  449.5 452.75    469  471.5
> flour.ybar<-.Last.value
```

EXAMPLES FROM CHAPTER 5   291

```
> flour.design<-fac.design(rep(2,6),
+ factor.names<-c("XF","XE","XD","XC","XB","XA"),
+ fraction = ~ XA:XB:XC:XD + XB:XC:XE:XF)
> flour.design
    XF  XE  XD  XC  XB  XA
 1  XF1 XE1 XD1 XC1 XB1 XA1
 2  XF2 XE2 XD1 XC1 XB1 XA1
 3  XF2 XE1 XD2 XC2 XB1 XA1
 4  XF1 XE2 XD2 XC2 XB1 XA1
 5  XF2 XE1 XD2 XC1 XB2 XA1
 6  XF1 XE2 XD2 XC1 XB2 XA1
 7  XF1 XE1 XD1 XC2 XB2 XA1
 8  XF2 XE2 XD1 XC2 XB2 XA1
 9  XF1 XE1 XD2 XC1 XB1 XA2
10  XF2 XE2 XD2 XC1 XB1 XA2
11  XF2 XE1 XD1 XC2 XB1 XA2
12  XF1 XE2 XD1 XC2 XB1 XA2
13  XF2 XE1 XD1 XC1 XB2 XA2
14  XF1 XE2 XD1 XC1 XB2 XA2
15  XF1 XE1 XD2 XC2 XB2 XA2
16  XF2 XE2 XD2 XC2 XB2 XA2

Fraction:  ~ XA:XB:XC:XD + XB:XC:XE:XF

> flour.df <- data.frame(flour.ybar, flour.design)

> flour.aov<-aov(flour.ybar~XA*XB*XC*XD*XE*XF,flour.df)
> summary(flour.aov)
          Df Sum of Sq   Mean Sq
      XA   1    53.473    53.473
      XB   1   752.816   752.816
      XC   1    89.066    89.066
      XD   1  1160.254  1160.254
      XE   1   412.598   412.598
      XF   1   230.660   230.660
   XA:XB   1   223.129   223.129
   XA:XC   1   382.691   382.691
   XB:XC   1   204.848   204.848
   XA:XE   1   412.598   412.598
   XB:XE   1   402.504   402.504
   XC:XE   1   387.598   387.598
   XD:XE   1   349.223   349.223
XA:XB:XE   1   692.348   692.348
XA:XC:XE   1   223.129   223.129

> flour.aov2<-aov(flour.y~flour.tmt+flour.day)
> summary(flour.aov2)
            Df Sum of Sq  Mean Sq  F Value     Pr(F)
flour.tmt   15   23907.7   1593.8  0.82706 0.6436536
flour.day    3  391397.8 130465.9 67.69952 0.0000000
Residuals   45   86721.0   1927.1

> model.tables(flour.aov,type="feffects")
```

```
Table of factorial effects
    XA     XB     XC     XD      XE     XF  XA:XB   XA:XC
5.1707 19.401 6.6733 24.086 -14.363 10.739 14.937 -19.563
 XB:XC  XD:XE XA:XB:XE XA:XC:XE XB:XC:XE XA:XE:XF XB:XE:XF
14.312 18.687   26.313  -14.937        0        0        0
```

C.6 Examples from Chapter 6

C.6.1 Split unit

The data for a split unit experiment are given in Table 6.8. The structure of this example is identical to the split unit example involving varieties of oats, originally given by Yates (1935), used as an illustration by Venables and Ripley (1999, Chapter 6.11) Their discussion of split unit experiments emphasizes their formal similarity to designs with more than one component of variance, such as discussed briefly in Section 6.5. From this point of view the subunits are nested within the whole units, and there is a special modelling operator A/B to represent factor B nested within factor A. Thus the result of

`aov(y ~ temp * prep + Error(reps/prep))`

is a list of aov objects, one of which is the whole unit analysis of variance and another is the subunit analysis of variance. The subunit analysis is implied by the model formula because the finest level analysis, in our case "within reps", is automatically computed. As with unbalanced data, `model.tables` cannot be used to obtain estimated standard errors, although it will work if the model statement is changed to omit the interaction term between preparation and temperature. Venables and Ripley (1999, Chapter 6.11) discuss the calculation of residuals and fitted values in models with more than one source of variation.

```
> y<-scan()
1: 30 34 29 35 41 26 37 38 33 36 42 36
13: 28 31 31 32 36 30 40 42 32 41 40 40
25: 31 35 32 37 40 34 41 39 39 40 44 45
37:
> prep<-factor(rep(1:3,12))
> temp<-factor(rep(rep(1:4,rep(3,4)),3))
> days<-factor(rep(1:3,rep(12,3)))
> split.design<-design(days,temp,prep)
> split.df<-data.frame(split.design,y)
> rm(y, prep, temp, days, split.design)
> split.df
  days temp prep  y
1    1    1    1 30
```

EXAMPLES FROM CHAPTER 6

```
2    1    1    2 34
3    1    1    3 29
4    1    2    1 35
        ...

> split.aov<-aov(y~temp*prep+Error(days/prep),split.df)
> summary(split.aov)

Error: days
          Df Sum of Sq Mean Sq F Value Pr(F)
Residuals  2    77.556  38.778

Error: prep %in% days
          Df Sum of Sq Mean Sq F Value   Pr(F)
     prep  2    128.39  64.194  7.0781 0.048537
Residuals  4     36.28   9.069

Error: Within
          Df Sum of Sq Mean Sq F Value   Pr(F)
     temp  3    434.08  144.69 36.427  0.000000
temp:prep  6     75.17   12.53  3.154  0.027109
Residuals 18     71.50    3.97

> model.tables(split.aov,type="mean")
Refitting model to allow projection

Tables of means
Grand mean

 36.028

 temp
    [,1]
1 31.222
2 34.556
3 37.889
4 40.444

 prep
    [,1]
1 35.667
2 38.500
3 33.917

 temp:prep
Dim 1 : temp
Dim 2 : prep
        1      2      3
1 29.667 33.333 30.667
2 34.667 39.000 30.000
3 39.333 39.667 34.667
4 39.000 42.000 40.333

#calculate errors by hand
```

```
#use whole plot error for prep;
#prep means are averaged over 12 obsns
> sqrt(2*9.06944/12)
[1] 1.229461

#use subplot error for temp;
#temp means are averaged over 9 obsns
> sqrt(2*3.9722/9)
[1] 0.9395271

#use subplot error for temp:prep;
#these means are averaged over 3 obsns
> sqrt(2*3.9722/3)
[1] 1.6273
```

C.6.2 Wafer experiment; Section 6.7.2

There are six controllable factors and one noise factor. The design is a split plot with the noise factor, over-etch time, the whole plot treatment. Each subplot is an orthogonal array of 18 runs with six factors each at three levels. Tables of such arrays are available within S-PLUS, using the command `oa.design`.

The F-value and p-value have been deleted from the output, as the main effects of the factors should be compared using the whole plot error, and the interactions of the factors with OE should be compared using the subplot error. These two error components are not provided using the split plot formula, as there is no replication of the whole plot treatment. One way to extract them is to specify the model with all estimable interactions, and pool the appropriate (higher order) ones to give an estimate of the residual mean square.

```
> elect1.design<-oa.design(rep(3,6))

> elect1.design
    A  B  C  D  E  G
1  A1 B1 C1 D1 E1 G1
2  A1 B2 C2 D2 E2 G2
3  A1 B3 C3 D3 E3 G3
4  A2 B1 C1 D2 E2 G3
5  A2 B2 C2 D3 E3 G1
6  A2 B3 C3 D1 E1 G2
7  A3 B1 C2 D1 E3 G2
8  A3 B2 C3 D2 E1 G3
9  A3 B3 C1 D3 E2 G1
10 A1 B1 C3 D3 E2 G2
11 A1 B2 C1 D1 E3 G3
12 A1 B3 C2 D2 E1 G1
13 A2 B1 C2 D3 E1 G3
```

EXAMPLES FROM CHAPTER 6

```
14 A2 B2 C3 D1 E2 G1
15 A2 B3 C1 D2 E3 G2
16 A3 B1 C3 D2 E3 G1
17 A3 B2 C1 D3 E1 G2
18 A3 B3 C2 D1 E2 G3

Orthogonal array design with 5 residual df.
Using columns 2, 3, 4, 5, 6, 7 from design oa.18.2p1x3p7

> OE<-factor(c(rep(1,18),rep(2,18)))

> elect.design<-design(elect1.design,OE)
Warning messages:
1: argument(s) 1 have 18 rows, will be replicated to 36 rows to
       match other arguments in: data.frame(elect1.design, OE)
2: Row names were wrong length, using default names in: data.f\
      rame(elect1.design, OE)

> elect.design
    A  B  C  D  E  G OE
1  A1 B1 C1 D1 E1 G1  1
2  A1 B2 C2 D2 E2 G2  1
3  A1 B3 C3 D3 E3 G3  1
4  A2 B1 C1 D2 E2 G3  1
   ...
33 A2 B3 C1 D2 E3 G2  2
34 A3 B1 C3 D2 E3 G1  2
35 A3 B2 C1 D3 E1 G2  2
36 A3 B3 C2 D1 E2 G3  2

> y<-scan()
1: 4750 5444 5802 6088 9000 5236 12960 5306 9370 4942
11: 5516 5084 4890 8334 10750 12508 5762 8692 5050 5884
21: 6152 6216 9390 5902 12660 5476 9812 5206 5614 5322
31: 5108 8744 10750 11778 6286 8920
37:

> elect.df<-data.frame(y,elect.design)
> rm(y, elect.design)
> elect.aov<-aov(y~(A+B+C+D+E+G)+OE+OE*(A+B+C+D+E+G),elect.df)
> summary(elect.aov)
         Df Sum of Sq   Mean Sq
      A   2  84082743  42041371
      B   2   6996828   3498414
      C   2   3289867   1644933
      D   2   5435943   2717971
      E   2  98895324  49447662
      G   2  28374240  14187120
     OE   1    408747    408747
   OE:A   2    112170     56085
   OE:B   2    245020    122510
   OE:C   2      5983      2991
   OE:D   2    159042     79521
   OE:E   2    272092    136046
```

```
       OE:G  2      13270      6635
Residuals 10     4461690    446169
```

```
> summary(elect.aov,split=list(A=list(1,2),B=list(1,2),
+                              C=list(1,2),D=list(1,2),
+                              E=list(1,2),G=list(1,2)))
               Df Sum of Sq  Mean Sq
            A   2  84082743 42041371
    A: Ctst 1   1  27396340 27396340
    A: Ctst 2   1  56686403 56686403
            B   2   6996828  3498414
    B: Ctst 1   1   5415000  5415000
    B: Ctst 2   1   1581828  1581828
            C   2   3289867  1644933
    C: Ctst 1   1   2275504  2275504
    C: Ctst 2   1   1014363  1014363
            D   2   5435943  2717971
    D: Ctst 1   1    130833   130833
    D: Ctst 2   1   5305110  5305110
            E   2  98895324 49447662
    E: Ctst 1   1  22971267 22971267
    E: Ctst 2   1  75924057 75924057
            G   2  28374240 14187120
    G: Ctst 1   1   2257067  2257067
    G: Ctst 2   1  26117174 26117174
           OE   1    408747   408747
         OE:A   2    112170    56085
 OE:A: Ctst 1   1       620      620
 OE:A: Ctst 2   1    111549   111549
         OE:B   2    245020   122510
 OE:B: Ctst 1   1    192963   192963
 OE:B: Ctst 2   1     52057    52057
         OE:C   2      5983     2991
 OE:C: Ctst 1   1      3220     3220
 OE:C: Ctst 2   1      2763     2763
         OE:D   2    159042    79521
 OE:D: Ctst 1   1     55681    55681
 OE:D: Ctst 2   1    103361   103361
         OE:E   2    272092   136046
 OE:E: Ctst 1   1      1734     1734
 OE:E: Ctst 2   1    270358   270358
         OE:G   2     13270     6635
 OE:G: Ctst 1   1     12331    12331
 OE:G: Ctst 2   1       939      939
    Residuals 10   4461690   446169
```

```
> summary(aov(y~A*B*C*D*E*G*OE,elect.df))
      Df Sum of Sq  Mean Sq
   A   2  84082743 42041371
   B   2   6996828  3498414
   C   2   3289867  1644933
   D   2   5435943  2717971
   E   2  98895324 49447662
   G   2  28374240 14187120
```

BIBLIOGRAPHIC NOTES   297

```
      OE  1    408747    408747
     A:B  2    229714    114857
     B:C  2   3001526   1500763
     B:E  1   1175056   1175056
    A:OE  2    112170     56085
    B:OE  2    245020    122510
    C:OE  2      5983      2991
    D:OE  2    159042     79521
    E:OE  2    272092    136046
    G:OE  2     13270      6635
  A:B:OE  2      2616      1308
  B:C:OE  2     49258     24629
  B:E:OE  1      3520      3520

> (229714+3001526+1175056)/5
[1] 881259.2
> (2616+49258+3520)/5
[1] 11078.8
```

C.7 Bibliographic notes

The definitive guide to statistical analysis with S-PLUS is Venables and Ripley (1999), now in its third edition. A detailed discussion of contrasts for fitting and partitioning sums of squares is given in Chapter 6.2, and analysis of structured designs is outlined in Chapter 6.7 and 6.8. Models with several components of variation are discussed in Chapter 6.11 and the latest release of S-PLUS includes a quite powerful method for fitting mixed effects models, lme. The software and sets of data for Venables and Ripley (1999) are available on the World Wide Web; a current list of sites is maintained at

{\tt http://www.stats.ox.ac.uk/pub/MASS3/sites.html}.

Chambers and Hastie (1992) is a useful reference for detailed understanding of the structure of data and models in S and has many examples of analysis of structured designs in Chapter 5. Their book refers to the S language, which is included in S-PLUS. Spector (1994) gives a readable introduction to both languages, with a number of useful programming tips. The manuals distributed with S-PLUS are useful for problems with the same structure as one of their examples: designed experiments are discussed in Chapters 13 through 15 of the S-PLUS 2000 Guide to Statistics, Vol I.

There are a number of S and S-PLUS functions available through the statlib site at Carnegie-Mellon University:

{\tt http://www.stat.cmu.edu}.

Of particular interest is the Designs archive at that site, which includes several programs for computing optimal designs, and the library of functions provided by F. Harrell (Harrell/hmisc in the S archive).

References

Abdelbasit, K.M. and Plackett, R.L. (1983). Experimental design for binary data. *J. Amer. Statist. Assoc.*, **78**, 90–98.

Armitage, P. (1975). *Sequential medical trials.* Oxford: Blackwell.

Aslett, R., Buck, R.J., Duvall, S.G., Sacks, J. and Welch, W.J. (1998). Circuit optimization via sequential computer experiments: design of an output buffer. *Appl. Statist.*, **47**, 31–48.

Atiqullah, M. and Cox, D.R. (1962). The use of control observations as an alternative to incomplete block designs. *J. R. Statist. Soc. B*, **24**, 464–471.

Atkinson, A.C. (1985). *Plots, transformation and regression.* Oxford University Press.

Atkinson, A.C. and Donev, A.N. (1992). *Optimal experimental designs.* Oxford University Press.

Atkinson, A.C. and Donev, A.N. (1996). Experimental designs optimally balanced against trend. *Technometrics*, **38**, 333–341.

Azaïs, J.-M., Monod, H. and Bailey, R.A. (1998). The influence of design on validity and efficiency of neighbour methods. *Biometrics*, **54**, 1374–1387.

Azzalini, A. and Cox, D.R. (1984). Two new tests associated with analysis of variance. *J. R. Statist. Soc. B*, **46**, 335–343.

Bailey, R.A. and Rowley, C.A. (1987). Valid randomization. *Proc. Roy. Soc. London*, A, **410**, 105–124.

Bartlett, M.S. (1933). The vector representation of a sample. *Proc. Camb. Phil. Soc.*, **30**, 327–340.

Bartlett, M.S. (1938). The approximate recovery of information from field experiments with large blocks. *J. Agric. Sci.*, **28**, 418–427.

Bartlett, M.S. (1978). Nearest neighbour models in the analysis of field experiments (with discussion). *J. R. Statist. Soc. B*, **40**, 147–170.

Bartlett, R.H., Roloff, D.W., Cornell, R.G., Andrews, A.F., Dillon, P.W. and Zwischenberger, J.B. (1985). Extracorporeal circulation in neonatal respiratory failure: A prospective randomized study. *Pediatrics*, **76**, 479–487.

Begg, C.B. (1990). On inferences from Wei's biased coin design for clinical trials (with discussion). *Biometrika*, **77**, 467–485

Besag, J. and Higdon, D. (1999). Bayesian analysis of agricultural field experiments (with discussion). *J. R. Statist. Soc.* B, **61**, 691–746.

Beveridge, W.V.I. (1952). *The art of scientific investigation.* London: Heinemann.

Biggers, J.D. and Heyner, R.R. (1961). Studies on the amino acid requirements of cartilaginous long bone rudiments *in vitro. J. Experimental Zoology*, **147**, 95–112.

Blackwell, D. and Hodges, J.L. (1957). Design for the control of selection bias. *Ann. Math. Statist.*, **28**, 449–460.

Blot, W.J. and 17 others (1993). Nutritional intervention trials in Linxian, China: supplementation with specific vitamin-mineral combinations, cancer incidence, and disease-specific mortality in the general population. *J. Nat. Cancer Inst.*, **85**, 1483–1492.

Booth, K.H.V. and Cox, D.R. (1962). Some systematic supersaturated designs. *Technometrics*, **4**, 489–495.

Bose, R.C. (1938). On the application of Galois fields to the problem of the construction of Hyper-Graeco Latin squares. *Sankhyā*, **3**, 323–338.

Bose, R.C. and Bush, K.A. (1952). Orthogonal arrays of strength two and three. *Ann. Math. Statist.*, **23**, 508–524.

Bose, R.C., Shrikhande, S.S. and Parker, E.T. (1960). Further results on the construction of mutually orthogonal Latin squares and the falsity of Euler's conjecture. *Canad. J. Math*, **12**, 189-203.

Box, G.E.P. and Draper, N.R. (1959). A basis for the selection of a response surface design. *J. Amer. Statist. Assoc.*, **54**, 622–654.

Box, G.E.P. and Draper, N.R. (1969). *Evolutionary operation.* New York: Wiley.

Box, G.E.P. and Hunter, J.S. (1957). Multi-factor experimental designs for exploring response surfaces. *Ann. Math. Statist.*, **28**, 195–241.

Box, G.E.P. and Lucas, H.L. (1959). Design of experiments in nonlinear situations. *Biometrika*, **46**, 77–90.

Box, G.E.P. and Wilson, K.B. (1951). On the experimental attainment of optimum conditions (with discussion). *J. R. Statist. Soc.*, B, **13**, 1–45.

Box, G.E.P., Hunter, W.G. and Hunter, J.S. (1978). *Statistics for experimenters.* New York: Wiley.

Brien, C.J. and Payne, R.W. (1999). Tiers, structure formulae and the analysis of complicated experiments. *J. R. Statist. Soc.* D, **48**, 41–52.

Carlin, B.P., Kadane, J. and Gelfand, A.E. (1998). Approaches for optimal sequential decision analysis in clinical trials. *Biometrics*, **54**, 964–975.

Carmichael, R.D. (1937). *Introduction to the theory of groups of finite order.* New York: Dover, 1956 reprint.

Chaloner, K. (1993). A note on optimal Bayesian design for nonlinear problems. *J. Statist. Plann. Inf.*, **37**, 229–235.

Chaloner, K. and Verdinelli, I. (1995). Bayesian experimental design: a review. *Statist. Sci.*, **10**, 273–304.

Chambers, J.M. and Hastie, T.J. (editors) (1992). *Statistical models in S*. Pacific Grove: Wadsworth & Brooks/Cole.

Chao, S.-C. and Shao, J. (1997). Statistical methods for two-sequence three-period cross-over trials with incomplete data. *Statistics in Medicine*, **16**, 1071–1039.

Cheng, C.-S., Martin, R.J. and Tang, B. (1998). Two-level factorial designs with extreme numbers of level changes. *Ann. Statist.*, **26**, 1522–1539.

Cheng, C.-S. and Mukerjee, R. (1998). Regular fractional factorial designs with minimum aberration and maximum estimation capacity. *Ann. Statist.*, **26**, 2289–2300.

Chernoff, H. (1953). Locally optimal designs for estimating parameters. *Ann. Math. Statist.*, **24**, 586–602.

Ciminera, J.L., Heyse, J.F., Nguyen, H. and Tukey, J.W. (1993). Tests for qualitative treatment by center interaction using a push-back procedure. *Statistics in Medicine*, **12**, 1033–1045.

Claringbold, P.J. (1955). Use of the simplex design in the study of the joint reaction of related hormones. *Biometrics*, **11**, 174–185.

Clarke, G.M. and Kempson, R.E. (1997). *Introduction to the design and analysis of experiments.* London: Arnold.

Cochran, W.G. and Cox, G.M. (1958). *Experimental designs.* Second edition. New York: Wiley.

Cook, R.D. and Weisberg, S. (1982). *Residuals and inference in regression.* London: Chapman & Hall.

Copas, J.B. (1973). Randomization models for the matched and unmatched 2 × 2 tables. *Biometrika*, **60**, 467–476.

Cornell, J.A. (1981). *Experiments with mixtures.* New York: Wiley.

Cornfield, J. (1978). Randomization by group: a formal analysis. *American J. Epidemiology*, **108**, 100–102.

Covey–Crump, P.A.K. and Silvey, S.D. (1970). Optimal regression designs with previous observations. *Biometrika*, **57**, 551–566.

Cox, D.R. (1951). Some systematic experimental designs. *Biometrika*, **38**, 310–315.

Cox, D.R. (1954). The design of an experiment in which certain treatment arrangements are inadmissible. *Biometrika*, **41**, 287–295.

Cox, D.R. (1957). The use of a concomitant variable in selecting an experimental design. *Biometrika*, **44**, 150–158.

Cox, D.R. (1958). *Planning of experiments.* New York: Wiley.

Cox, D.R. (1971). A note on polynomial response functions for mixtures. *Biometrika*, **58**, 155–159.

Cox, D.R. (1982). Randomization and concomitant variables in the design of experiments. In *Statistics and Probability: Essays in honor of C.R. Rao.* Editors G. Kallianpur, P.R. Krishnaiah and J.K. Ghosh.

Amsterdam: North Holland, pp. 197–202.

Cox, D.R. (1984a). Interaction (with discussion). *Int. Statist. Rev.*, **52**, 1–31.

Cox, D.R. (1984b). Effective degrees of freedom and the likelihood ratio test. *Biometrika*, **71**, 487–493.

Cox, D.R. (1992). Causality: some statistical aspects. *J. R. Statist. Soc.*, A, **155**, 291–301.

Cox, D.R. and Hinkley, D.V. (1974). *Theoretical statistics.* London: Chapman & Hall.

Cox, D.R. and McCullagh, P. (1982). Some aspects of analysis of covariance (with discussion). *Biometrics*, **38**, 541–561.

Cox, D.R. and Snell, E.J. (1981). *Applied statistics.* London: Chapman & Hall.

Cox, D.R. and Wermuth, N. (1996). *Multivariate dependencies.* London: Chapman & Hall.

Daniel, C. (1959). Use of half normal plot in interpreting factorial two-level experiments. *Technometrics*, **1**, 311–341.

Daniel, C. (1994). Factorial one-factor-at-a-time experiments. *American Statistician*, **48**, 132–135.

Davies, O.L. (editor) (1956). *Design and analysis of industrial experiments.* 2nd ed. Edinburgh: Oliver & Boyd.

Dawid, A.P. (2000). Causality without counterfactuals (with discussion). *J. Amer. Statist. Assoc.*, **95**, to appear.

Dawid, A.P. and Sebastiani, P. (1999). Coherent dispersion criteria for optimal experimental design. *Ann. Statist.*, **27**, 65–81.

Dean, A. and Voss, D. (1999). *Design and analysis of experiments.* New York: Springer.

Denés, J. and Keedwell, A.D. (1974). *Latin squares and their applications.* London: English Universities Press.

Desu, M.M. and Raghavarao, D. (1990). *Sample size methodology.* New York: Academic Press.

Dey, A. and Mukerjee, R. (1999). *Fractional factorial plans.* New York: Wiley.

Donnelly, C.A. and Ferguson, N.M. (1999). *Statistical aspects of BSE and vCJD.* London: Chapman & Hall.

Draper, N. and Smith, H. (1998). *Applied regression analysis.* 3rd edition. New York: Wiley.

Easton, D.F., Peto, J. and Babiker, A.G. (1991). Floating absolute risk: alternative to relative risk in survival and case-control analysis avoiding and arbitrary reference group. *Statistics in Medicine*, **10**, 1025–1035.

Elfving, G. (1952). Optimum allocation in linear regression theory. *Ann. Math. Statist.*, **23**, 255–262.

Elfving, G. (1959). Design of linear experiments. In *Cramër Festschrift* volume, ed. U. Grenander, pp.58–74. New York: Wiley.

Fang, K.-T. and Wang, Y. (1993). *Number-theoretic methods in statistics.* London: Chapman & Hall.

Fang, K.-T., Wang, Y. and Bentler, P.M. (1994). Some applications of number-theoretic methods in statistics. *Statist. Sci.*, **9**, 416–428.

Farewell, V.T. and Herzberg, A.M. (2000). Plaid designs for the evaluation of training for medical practitioners. To appear.

Fearn, T. (1992). Box-Cox transformations and the Taguchi method: an alternative analysis of a Taguchi case study. *Appl. Statist.*, **41**, 553–559.

Fedorov, V.V. (1972). *Theory of optimal experiments.* (English translation from earlier Russian edition). New York: Academic Press.

Fedorov, V.V. and Hackl, P. (1997). *Model oriented design of experiments.* New York: Springer.

Finney, D.J. (1945a). Some orthogonal properties of the 4×2 and 6×6 Latin squares. *Ann. Eugenics*, **12**, 213–217.

Finney, D.J. (1945b). The fractional replication of factorial arrangements. *Ann. Eugenics*, **12**, 283–290.

Firth, D. and Menezes, R. (2000). Quasi-variances for comparing groups: control relative not absolute error. To appear.

Fisher, R.A. (1926). The arrangement of field experiments. *J. Ministry of Agric.*, **33**, 503–513.

Fisher, R.A. (1935). *Design of experiments.* Edinburgh: Oliver & Boyd.

Fisher, R.A. and Mackenzie, W.A. (1923). Studies in crop variation II. The manurial response of different potato varieties. *J. Agric. Sci.*, **13**, 311–320.

Flournoy, N., Rosenberger, W.F. and Wong, W.K. (1998) (eds). *New developments and applications in experimental design.* Hayward: Institute of Mathematical Statistics.

Fries, A. and Hunter, W.G. (1980). Minimum aberration in 2^{k-p} designs. *Technometrics*, **222**, 601–608.

Gail, M. and Simon, R. (1985). Testing for qualitative interaction between treatment effects and patient subsets. *Biometrics*, **41**, 361–372.

Gilmour, A.R., Cullis, B.R. and Verbyla, A.P. (1997). Accounting for natural and extraneous variation in the analysis of field experiments. *J. Agric. Bio. Environ. Statist.*, **2**, 269–273.

Gilmour, S.G. and Ringrose, T.J. (1999). Controlling processes in food technology by simplifying the canonical form of fitted response surfaces. *Appl. Statist.*, **48**, 91–102.

Ginsberg, E.S., Mello, N.K., Mendelson, J.H., Barbieri, R.L., Teoh, S.K., Rothman, M., Goa, X. and Sholar, J.W. (1996). Effects of alcohol ingestion on œstrogens in postmenopausal women. *J. Amer. Med. Assoc.* **276**, 1747–1751.

Goetghebeur, E. and Houwelingen, H.C. van (1998)(eds). Special issue on noncompliance in clinical trials. *Statistics in Medicine*, **17**, 247–390.

Good, I.J. (1958). The interaction algorithm and practical Fourier analysis. *J. R. Statist. Soc.*, B, **20**, 361–372.

Grundy, P.M. and Healy, M.J.R. (1950). Restricted randomization and quasi-Latin squares. *J. R. Statist. Soc.*, B, **12**, 286–291.

Guyatt, G., Sackett, D., Taylor, D.W., Chong, J., Roberts, R. and Pugsley, S. (1986). Determining optimal therapy-randomized trials in individual patients. *New England J. Medicine*, **314**, 889–892.

Hald, A. (1948). *The decomposition of a series of observations.* Copenhagen: Gads Forlag.

Hald, A. (1998). *A history of mathematical statistics.* New York: Wiley.

Hartley, H.O. and Smith, C.A.B. (1948). The construction of Youden squares. *J. R. Statist. Soc.*, B, **10**, 262–263.

Hedayat, A.S., Sloane, N.J.A. and Stufken, J. (1999). *Orthogonal arrays: theory and applications.* New York: Springer.

Heise, M.A. and Myers, R.H. (1996). Optimal designs for bivariate logistic regression. *Biometrics*, **52**, 613–624.

Herzberg, A.M. (1967). The behaviour of the variance function of the difference between two estimated responses. *J. R. Statist. Soc.*, B, **29**, 174–179.

Herzberg, A.M. and Cox, D.R. (1969). Recent work on the design of experiments: a bibliography and a review. *J. R. Statist. Soc.*, A, **132**, 29–67.

Hill, R. (1986). *A first course in coding theory.* Oxford University Press.

Hinkelman, K. and Kempthorne, O. (1994). *Design and analysis of experiments.* New York: Wiley.

Holland, P.W. (1986). Statistics and causal inference (with discussion). *J. Amer. Statist. Assoc.*, **81**, 945–970.

Huang, Y.-C. and Wong, W.-K. (1998). Sequential considerations of multiple-objective optimal designs. *Biometrics*, **54**, 1388–1397.

Hurrion, R.D. and Birgil, S. (1999). A comparison of factorial and random experimental design methods for the development of regression and neural simulation metamodels. *J. Operat. Res. Soc.*, **50**, 1018–1033.

Jennison, C. and Turnbull, B.W. (2000). *Group sequential methods with applications to clinical trials.* London: Chapman & Hall.

John, J.A. and Quenouille, M.H. (1977). *Experiments: design and analysis.* London: Griffin.

John, J.A., Russell, K.G., Williams, E.R. and Whitaker, D. (1999). Resolvable designs with unequal block sizes. *Austr. and NZ J. Statist.*, **41**, 111–116.

John, J.A. and Williams, E.R. (1995). *Cyclic and computer generated designs.* 2nd edition. London: Chapman & Hall.

REFERENCES

John, P.W.M. (1971). *Statistical design and analysis of experiments.* New York: Macmillan.

Johnson, T. (1998). Clinical trials in psychiatry: background and statistical perspective. *Statist. Methods in Medical Res.*, **7**, 209–234.

Jones, B. and Kenward, M.G. (1989). *Design and analysis of crossover trials.* London: Chapman & Hall.

Kempthorne, O. (1952). *Design of experiments.* New York: Wiley.

Kiefer, J. (1958). On the nonrandomized optimality and randomized nonoptimality of symmetrical designs. *Ann. Math. Statist.*, **29**, 675–699.

Kiefer, J. (1959). Optimum experimental design (with discussion). *J. R. Statist. Soc.*, B, **21**, 272–319.

Kiefer, J. (1975). Optimal design: variation in structure and performance under change of criterion. *Biometrika*, **62**, 277–288.

Kiefer, J. (1985). *Collected papers.* eds. L. Brown, I. Olkin, J. Sacks and H.P. Wynn. New York: Springer.

Kiefer, J. and Wolfowitz, J. (1959). Optimal designs in regression problems. *Ann. Math. Statist.*, **30**, 271–294.

Kruskal, W.H. (1961). The coordinate-free approach to Gauss-Markov estimation, and its application to missing and extra observations. *Proc. 4th Berkeley Symposium*, **1**, 435–451.

Lauritzen, S.L. (2000). Causal inference from graphical models. In *Complex stochastic systems.* C. Klüppelberg, O.E. Barndorff-Nielsen and D.R. Cox, editors. London: Chapman & Hall/CRC.

Leber, P.D. and Davis, C.S. (1998). Threats to the validity of clinical trials employing enrichment strategies for sample selection. *Controlled Clinical Trials*, **19**, 178–187.

Lehmann, E.L. (1975). *Nonparametrics: statistical methods based on ranks.* San Francisco: Holden-Day.

Lindley, D.V. (1956). On the measure of information provided by an experiment. *Ann. Math. Statist.*, **27**, 986–1005.

Logothetis, N. (1990). Box-Cox transformations and the Taguchi method. *Appl. Statist.*, **39**, 31–48.

Logothetis, N. and Wynn, H.P. (1989). *Quality through design.* Oxford University Press.

McCullagh, P. (2000). Invariance and factorial models (with discussion). *J. R. Statist. Soc.*, B, **62**, 209–256.

McCullagh, P. and Nelder, J.A. (1989). *Generalized linear models.* 2nd edition. London: Chapman & Hall.

McKay, M.D., Beckman, R.J. and Conover, W.J. (1979). A comparison of three methods for selecting values of input variables in the analysis of output from a computer code. *Technometrics*, **21**, 239–245.

Manly, B.J.F. (1997). *Randomization, bootstrap and Monte Carlo methods in biology.* London: Chapman & Hall.

Mehrabi, Y. and Matthews, J.N.S. (1998). Implementable Bayesian designs for limiting dilution assays. *Biometrics*, **54**, 1398–1406.

Mesenbrink, P., Lu, J-C., McKenzie, R. and Taheri, J. (1994). Characterization and optimization of a wave-soldering process. *J. Amer. Statist. Assoc.*, **89**, 1209–1217.

Meyer, R.D., Steinberg, D.M. and Box, G.E.P. (1996). Follow up designs to resolve confounding in multifactorial experiments. *Technometrics*, **38**, 303–318.

Monod, H., Azaïs, J.-M. and Bailey, R.A. (1996). Valid randomization for the first difference analysis. *Austr. J. Statist.*, **38**, 91–106.

Montgomery, D.C. (1997). *Design and analysis of experiments.* 4th edition. New York: Wiley.

Montgomery, D.C. (1999). Experimental design for product and process design and development (with comments). *J. R. Statist. Soc.*, D, **48**, 159–177.

Nair, V.J. (editor) (1992). Taguchi's parameter design: a panel discussion. *Technometrics*, **34**, 127–161.

Neiderreiter, H. (1992). *Random number generation and quasi-Monte Carlo methods.* Philiadelphia: SIAM.

Nelder, J.A. (1965a). The analysis of experiments with orthogonal block structure. I Block structure and the null analysis of variance. *Proc. Roy. Soc. London*, A, **283**, 147–162.

Nelder, J.A. (1965b). The analysis of experiments with orthogonal block structure. II Treatment structure and the general analysis of variance. *Proc. Roy. Soc. London*, A, **283**, 163–178.

Newcombe, R.G. (1996). Sequentially balanced three-squares cross-over designs. *Statistics in Medicine*, **15**, 2143–2147.

Neyman, J. (1923). On the application of probability theory to agricultural experiments. Essay on principles. *Roczniki Nauk Rolniczych*, **10**, 1–51 (in Polish). English translation of Section 9 by D.M. Dabrowska and T.P. Speed (1990), *Statist. Sci.*, **9**, 465–480.

Olguin, J. and Fearn, T. (1997). A new look at half-normal plots for assessing the significance of contrasts for unreplicated factorials. *Appl. Statist.*, **46**, 449–462.

Owen, A. (1992). Orthogonal arrays for computer experiments, integration, and visualization. *Statist. Sinica*, **2**, 459–452.

Owen, A. (1993). A central limit theorem for Latin hypercube sampling. *J. R. Statist. Soc.*, B, **54**, 541–551.

Papadakis, J.S. (1937). Méthods statistique poure des expériences sur champ. *Bull. Inst. Amér. Plantes à Salonique*, No. 23.

Patterson, H.D. and Williams, E.R. (1976). A new class of resolvable incomplete block designs. *Biometrika*, **63**, 83–92.

Pearce, S.C. (1970). The efficiency of block designs in general. *Biometrika* **57**, 339–346.

Pearl, J. (2000). *Causality: models, reasoning and inference*. Cambridge: Cambridge University Press.

Pearson, E.S. (1947). The choice of statistical tests illustrated on the interpretation of data classed in a 2 × 2 table. *Biometrika*, **34**, 139–167.

Piantadosi, S. (1997). *Clinical trials*. New York: Wiley.

Pistone, G. and Wynn, H.P. (1996). Generalised confounding with Gröbner bases. *Biometrika*, **83**, 653–666.

Pistone, G., Riccomagno, E. and Wynn, H.P. (2000). *Algebraic statistics*. London: Chapman & Hall/CRC.

Pitman, E.J.G. (1937). Significance tests which may be applied to samples from any populations: III The analysis of variance test. *Biometrika*, **29**, 322–335.

Plackett, R.L. and Burman, J.P. (1945). The design of optimum multifactorial experiments. *Biometrika*, **33**, 305–325.

Preece, A.W., Iwi, G., Davies-Smith, A., Wesnes, K., Butler, S., Lim, E. and Varney, A. (1999). Effect of 915-MHz simulated mobile phone signal on cognitive function in man. *Int. J. Radiation Biology*, **75**, 447–456.

Preece, D.A. (1983). Latin squares, Latin cubes, Latin rectangles, etc. In *Encyclopedia of statistical sciences, Vol.4*. S. Kotz and N.L. Johnson, eds, 504–510.

Preece, D.A. (1988). Semi-Latin squares. In *Encyclopedia of statistical sciences, Vol.8*. S. Kotz and N.L. Johnson, eds, 359–361.

Pukelsheim, F. (1993). *Optimal design of experiments*. New York: Wiley.

Quenouille, M.H. (1953). *The design and analysis of experiments*. London: Griffin.

Raghavarao, D. (1971). *Construction and combinatorial problems in design of experiments*. New York: Wiley.

Raghavarao, D. and Zhou, B. (1997). A method of constructing 3-designs. *Utilitas Mathematica*, **52**, 91–96.

Raghavarao, D. and Zhou, B. (1998). Universal optimality of UE 3-designs for a competing effects model. *Comm. Statist.–Theory Meth.*, **27**, 153–164.

Rao, C. R. (1947). Factorial experiments derivable from combinatorial arrangements of arrays. *Suppl. J. R. Statist. Soc.*, **9**, 128–139.

Redelmeier, D. and Tibshirani, R. (1997a). Association between cellular phones and car collisions. *N. England J. Med.*, **336**, 453–458.

Redelmeier, D. and Tibshirani, R.J. (1997b). Is using a cell phone like driving drunk? *Chance*, **10**, 5–9.

Reeves, G.K. (1991). Estimation of contrast variances in linear models. *Biometrika*, **78**, 7–14.

Ridout, M.S. (1989). Summarizing the results of fitting generalized linear models from designed experiments. In *Statistical modelling: Proceed-*

ings of *GLIM89*. A. Decarli et al. editors, pp. 262–269. New York: Springer.

Robbins, H. and Monro, S. (1951). A stochastic approximation method. *Ann. Math. Statist.*, **22**, 400–407.

Rosenbaum, P.R. (1987). The role of a second control group in an observational study (with discussion). *Statist. Sci.*, **2**, 292–316.

Rosenbaum, P.R. (1999). Blocking in compound dispersion experiments. *Technometrics*, **41**, 125–134.

Rosenberger, W.F. and Grill, S.E. (1997). A sequential design for psychophysical experiments: an application to estimating timing of sensory events. *Statistics in Medicine*, **16**, 2245–2260.

Rubin, D.B. (1974). Estimating causal effects of treatments in randomized and nonrandomized studies. *J. Educ. Psychol.*, **66**, 688–701.

Sacks, J., Welch, W.J., Mitchell, T.J. and Wynn, H.P. (1989). Design and analysis of computer experiments (with discussion). *Statist. Sci.*, **4**, 409–436.

Sattherthwaite, F. (1958). Random balanced experimentation. *Technometrics*, **1**, 111–137.

Scheffé, H. (1958). Experiments with mixtures (with discussion). *J.R. Statist. Soc.*, B, **20**, 344–360.

Scheffé, H. (1959). *Analysis of variance.* New York: Wiley.

Senn, S.J. (1993). *Cross-over trials in clinical research.* Chichester: Wiley.

Shah, K.R. and Sinha, B.K. (1989). *Theory of optimal designs.* Berlin: Springer.

Silvey, S.D. (1980). *Optimal design.* London: Chapman & Hall.

Singer, B.H. and Pincus, S. (1998). Irregular arrays and randomization. *Proc. Nat. Acad. Sci. USA*, **95**, 1363–1368.

Smith, K. (1918). On the standard deviation of adjusted and interpolated values of an observed polynomial function and its constants, and the guidance they give towards a proper choice of the distribution of observations. *Biometrika*, **12**, 1–85.

Spector, P. (1994). *An introduction to S and S-Plus.* Belmont: Duxbury.

Speed, T.P. (1987). What is an analysis of variance? (with discussion). *Ann. Statist.*, **15**, 885–941.

Stein, M. (1987). Large sample properties of simulations using Latin hypercube sampling. *Technometrics*, **29**, 143–151.

Stigler, S.M. (1986). *The history of statistics.* Cambridge, Mass: Harvard University Press.

Street, A.P. and Street, D.J. (1987). *Combinatorics of experimental design.* Oxford University Press.

Tang, B. (1993). Orthogonal array-based Latin hypercubes. *J. Amer. Statist. Assoc.*, **88**, 1392–1397.

Thompson, M. E. (1997). *Theory of sample surveys.* London: Chapman

& Hall.

Tsai, P.W., Gilmour, S.G., and Mead, R. (1996). An alternative analysis of Logothetis's plasma etching data. Letter to the editor. *Appl. Statist.*, **45**, 498–503.

Tuck, M.G., Lewis, S.M. and Cottrell, J.I.L. (1993). Response surface methodology and Taguchi: a quality improvement study from the milling industry. *Appl. Statist.*, **42**, 671–681.

Tukey, J.W. (1949). One degree of freedom for non-additivity. *Biometrics*, **5**, 232–242.

UK Collaborative ECMO Trial Group. (1986). UK collaborative randomised trial of neonatal extracorporeal membrane oxygenation. *Lancet*, **249**, 1213–1217.

Vaart, A.W. van der (1998). *Asymptotic statistics.* Cambridge: Cambridge University Press.

Vajda, S. (1967a). *Patterns and configurations in finite spaces.* London: Griffin.

Vajda, S. (1967b). *The mathematics of experimental design; incomplete block designs and Latin squares.* London: Griffin.

Venables, W.M. and Ripley, B.D. (1999). *Modern applied statistics with S-PLUS.* 3rd ed. Berlin: Springer.

Wald, A. (1947). *Sequential analysis.* New York: Wiley.

Wang, J.C. and Wu, C.F.J. (1991). An approach to the construction of asymmetric orthogonal arrays. *J. Amer. Statist. Assoc.*, **86**, 450–456.

Wang, Y.-G. and Leung, D. H.-Y. (1998). An optimal design for screening trials. *Biometrics*, **54**, 243–250.

Ware, J.H. (1989). Investigating therapies of potentially great benefit: ECMO (with discussion). *Statist. Sci.*, **4**, 298–340.

Wei, L.J. (1988). Exact two-sample permutation tests based on the randomized play-the-winner rule. *Biometrika*, **75**, 603–606.

Welch, B.L. (1937). On the z test in randomized blocks and Latin squares. *Biometrika*, **29**, 21–52.

Wetherill, G.B. and Glazebrook, K.D. (1986). *Sequential methods in statistics.* 3rd edition. London: Chapman & Hall.

Whitehead, J. (1997). *The design and analysis of sequential medical trials.* 2nd edition. Chichester: Wiley.

Williams, E.J. (1949). Experimental designs balanced for the estimation of residual effects of treatments. *Australian J. Sci. Res.*, A, **2**, 149–168.

Williams, E.J. (1950). Experimental designs balanced for pairs of residual effects. *Australian J. Sci. Res.*, A, **3**, 351–363.

Williams, R.M. (1952). Experimental designs for serially correlated observations. *Biometrika*, **39**, 151–167.

Wilson, E.B. (1952). *Introduction to scientific research.* New York: McGraw Hill.

Wynn, H.P. (1970). The sequential generation of D-optimum experimental designs. *Ann. Statist.*, **5**, 1655–1664.

Yates, F. (1935). Complex experiments (with discussion). *Suppl. J. R. Statist. Soc.*, **2**, 181–247.

Yates, F. (1936). A new method of arranging variety trials involving a large number of varieties. *J. Agric. Sci.*, **26**, 424–455.

Yates, F. (1937). *The design and analysis of factorial experiments.* Technical communication **35**. Harpenden: Imperial Bureau of Soil Science.

Yates, F. (1951a). Bases logiques de la planification des expériences. *Ann. Inst. H. Poincaré*, **12**, 97–112.

Yates, F. (1951b). Quelques développements modernes dans la planification des expériences. *Ann. Inst. H. Poincaré*, **12**, 113–130.

Yates, F. (1952). Principles governing the amount of experimentation in developmental work. *Nature*, **170**, 138–140.

Youden, W.J. (1956). Randomization and experimentation (abstract). *Ann. Math. Statist.*, **27**, 1185–1186.

List of tables

3.1	Strength index of cotton	51
3.2	Analysis of variance for strength index	51
4.1	5 × 5 Graeco-Latin square	69
4.2	Complete orthogonal set of 5 × 5 Latin squares	70
4.3	Balanced incomplete block designs	72
4.4	Two special incomplete block designs	73
4.5	Youden square	75
4.6	Intrablock analysis of variance	78
4.7	Interblock analysis of variance	80
4.8	Analysis of variance, general incomplete block design	82
4.9	Log dry weight of chick bones	84
4.10	Treatment means, log dry weight	84
4.11	Analysis of variance, log dry weight	84
4.12	Estimates of treatment effects	85
4.13	Expansion index of pastry dough	86
4.14	Unadjusted and adjusted treatment means	86
4.15	Example of intrablock analysis	87
5.1	Weights of chicks	102
5.2	Mean weights of chicks	103
5.3	Two factor analysis of variance	108
5.4	Analysis of variance, weights of chicks	109
5.5	Decomposition of treatment sum of squares	110
5.6	Analysis of variance, 2^2 factorial	113
5.7	Analysis of variance, 2^k factorial	116
5.8	Treatment factors, nutrition and cancer	120
5.9	Data, nutrition and cancer	121
5.10	Estimated effects, nutrition and cancer	121
5.11	Data, Exercise 5.6	125
5.12	Contrasts, Exercise 5.6	125

6.1	Example, double confounding	130
6.2	Degrees of freedom, double confounding	131
6.3	Two orthogonal Latin squares	132
6.4	Estimation of the main effect	133
6.5	1/3 fraction, degrees of freedom	135
6.6	Asymmetric orthogonal array	138
6.7	Supersaturated design	139
6.8	Example, split-plot analysis	143
6.9	Data, tensile strength of paper	144
6.10	Analysis, tensile strength of paper	145
6.11	Table of means, tensile strength of paper	145
6.12	Analysis of variance for a replicated factorial	147
6.13	Analysis of variance with a random effect	149
6.14	Analysis of variance, quality/quantity interaction	154
6.15	Design, electronics example	159
6.16	Data, electronics example	160
6.17	Analysis of variance, electronics example	161
6.18	Example, factorial treatment structure for incomplete block design	167
8.1	Treatment allocation: biased coin design	207
8.2	3×3 lattice squares for nine treatments	217
8.3	4×4 Latin square	218
A.1	Stagewise analysis of variance table	235
A.2	Analysis of variance, nested and crossed	241
A.3	Analysis of variance for $Y_{abc;j}$	242
A.4	Analysis of variance for $Y_{(a;j)bc}$	242
B.1	Construction of two orthogonal 5×5 Latin squares	257

Author index

Abdelbasit, K.M. 188, 299
Andrews, A.F. 299
Armitage, P. 220, 299
Aslett, R. 188, 299
Atiqullah, M. 97, 299
Atkinson, A.C. 187, 220, 245, 246, 299
Azaïs, J.-M. 220, 299, 306
Azzalini, A. 122, 299
Babiker, A.G. 302
Bailey, R.A. 37, 220, 264, 299, 306
Barbieri, R.L. 303
Bartlett, M.S. 220, 244, 299
Bartlett, R.H. 223, 299
Beckman, R.J. 187, 305
Begg, C.J. 224, 299
Bentler, P.M. 188, 303
Besag, J. 220, 299
Beveridge, W.V.I. 15, 300
Biggers, J.D. 83, 300
Birgil, S. 162, 304
Blackwell, D. 223, 300
Blot, W.J. 120, 121, 122, 288, 300
Booth, K.H.V. 162, 300
Bose, R.C. 70, 162, 244, 264, 300
Box, G.E.P. 14, 15, 122, 123, 157, 162, 163, 165, 166, 186, 187, 300, 306
Brien, C.J. 163, 245, 300
Buck, R.J. 299
Burman, J.P. 122, 162, 305
Bush, K.A. 162, 300
Butler, S. 307
Carlin, B.P. 220, 300
Carmichael, R.D. 264, 300

Chaloner, K. 187, 300
Chambers, J.M. 270, 271, 273, 297, 301
Chao, S.-C. 95, 301
Cheng, C.-S. 125, 126, 301
Chernoff, H. 186, 301
Chong, J. 304
Ciminera, J.L. 122, 301
Claringbold, P.J. 163, 301
Clarke, G.M. 15, 301
Cochran, W.G. 50, 51, 62, 95, 162, 273, 301
Conover, W.J. 187, 305
Cook, R.D. 245, 246, 301
Copas, J.B. 38, 301
Cornell, J.A. 163, 301
Cornell, R.G. 299
Cornfield, J. 39, 301
Cottrell, J.I.L. 124, 125, 289, 309
Covey-Crump, P.A.K. 187, 299
Cox, D.R. 15, 37, 38, 62, 64, 83, 97, 101, 122, 162, 163, 190, 220, 245, 299–304
Cox, G.M. 50, 51, 62, 95, 162, 273, 301
Cullis, B.R. 220, 303
Daniel, C. 122, 124, 163, 302
Davies, O.L. 123, 302
Davies-Smith, A. 307
Davis, C.S. 220, 305
Dawid, A.P. 15, 187, 302
Dean, A. 15, 62, 166, 302
Denés, J. 95, 302
Desu, M.M. 219, 302
Dey, A. 125, 162, 187, 302

Dillon, P.W. 299
Donev, A.N. 187, 220, 299
Donnelly, C.M. 188, 302
Draper, N.R. 163, 187, 245, 300, 302
Duvall, S.G. 299
Easton, D.F. 245, 302
Elfving, G. 186, 302
Fang, K.-T. 188, 190, 303
Farewell, V.T. 165, 303
Fearn, T. 122, 163, 303, 306
Fedorov, V.V. 186, 187, 303
Ferguson, N.M. 188, 302
Finney, D.J. 96, 122, 264, 303
Firth, D. 245, 303
Fisher, R.A. 14, 36, 62, 95, 122, 166, 186, 244, 245, 303
Flournoy, N. 187, 189, 303
Fries, A. 124, 303
Gail, M.H. 122, 303
Gelfand, A.E. 220, 300
Gilmour, A.R. 220, 303
Gilmour, S.G. 86 , 166, 167, 303, 309
Ginsberg, E.S. 16, 303
Glazebrook, K.D. 222, 309
Goa, X. 303
Goetghebeur, E. 15, 303
Good, I.J. 123, 304
Grill, S.E. 220, 308
Grundy, P.M. 264, 304
Guyatt, G. 95, 304
Hackl, P. 187, 303
Hald, A. 220, 244, 304
Harrell, F. 298
Hartley, H.O. 95, 304
Hastie, T.J. 270, 271, 273, 297, 301
Healy, M.J.R. 264, 304
Hedayat, A.S. 162, 264, 304
Heise, M.A. 188, 304
Herzberg, A.M. 15, 165, 166, 303, 304
Heyner, R.R. 83, 300

Heyse, J.F. 301
Higdon, J. 220, 299
Hill, R. 122, 264, 304
Hinkelman, K. 15, 37, 62, 304
Hinkley, D.V. 38, 62, 302
Hodges, J.L. 223, 300
Holland, P.W. 15, 37, 304
Houwelingen, H.C. van 15, 303
Huang, Y.-C. 304
Hunter, J.S. 15, 122, 162, 163, 165, 166, 300
Hunter, W.G. 15, 122, 162, 300, 303
Hurrion, R.D. 162, 304
Iwi, G. 307
Jennison, C. 220, 304
John, J.A. 59, 95, 97, 101, 264, 304
John, P.W.M. 95, 305
Johnson, T. 39, 305
Jones, B. 95, 305
Kadane, J. 220, 300
Keedwell, A.D. 95, 302
Kempson, R.E. 15, 301
Kempthorne, O. 15, 37, 62, 162, 304, 305
Kenward, M.G. 95, 305
Kiefer, J. 186, 188, 220, 302, 305
Kirkman, T.P. 96
Kruskal, W.H. 244, 305
Lauritzen, S.L. 15, 305
Leber, P.D. 220, 305
Lehmann, E.L. 39, 305
Leung, D.H.-Y. 220, 309
Lewis, S.M. 124, 125, 289, 309
Lim, E. 307
Lindley, D.V. 187, 305
Logothetis, N. 158, 163, 305
Lu, J.-C. 126, 306
Lucas, H.L. 186, 300
Mackenzie, W.A. 245, 301
Manly, B.J.F. 37, 305
Matthews, J.N.S. 187, 306

AUTHOR INDEX

McCullagh, P. 37, 38, 122, 245, 302, 305
McKay, M.D. 187, 305
McKenzie, R. 126, 306
Martin, R.J. 126, 301
Mead, R. 309
Mehrabi, Y. 187, 306
Menezes, R. 245, 303
Mello, N.K. 303
Mendelson, J.H. 303
Mesenbrink, P. 126, 306
Meyer, R.D. 163, 306
Mitchell, T.J. 187, 308
Monod, H. 220, 299, 306
Monro, S. 222, 308
Montgomery, D.C. 15, 62, 143, 144, 306
Mukerjee, R. 125, 162, 187, 300, 302
Myers, R.H. 188, 304
Nair, V.J. 163, 306
Neiderreiter, H. 188, 306
Nelder, J.A. 38, 163, 245, 305, 306
Newcombe, R.G. 95, 306
Neyman, J. 37, 220, 306
Nguyen, H. 301
Olguin, J. 122, 306
Owen, A. 188, 190, 306
Papadakis, J.S. 220, 306
Parker, E.T. 70, 264, 300
Patterson, H.D. 97, 306
Payne, R.W. 163, 245, 300
Pearce, S.C. 95, 306
Pearl, J. 15, 307
Pearson, E.S. 38, 307
Peto, J. 302
Piantadosi, S. 15, 122, 307
Pincus, S. 37, 308
Pistone, G. 167, 168, 307
Pitman, E.J.G. 37, 307
Plackett, R.L. 122, 162, 188, 299, 307
Preece, A.W. 17, 307
Preece, D.A. 95, 307

Pugsley, S. 304
Pukelsheim, F. 187, 307
Quenouille, M.H. 101, 163, 304, 307
Raghavarao, D. 95, 97, 219, 302, 307
Rao, C.R. 162, 307
Redelmeier, D. 15, 16, 307
Reeves, G.K. 245, 307
Riccomagno, E. 168, 307
Ridout, M. 245, 307
Ringrose, T.J. 86, 166, 167, 303
Ripley, B.D. 268, 272, 273, 292, 297, 309
Robbins, H. 222, 308
Roberts, R. 304
Roloff, D.W. 299
Rosenbaum, P.R. 17, 37, 163, 308
Rosenberger, W.F. 187, 198, 220, 303, 308
Rothman, M. 303
Rowley, C.A. 37, 264, 299
Rubin, D.B. 15, 308
Russell, K.G. 304
Sackett, D. 304
Sacks, J. 187, 299, 308
Satterthwaite, F. 162, 308
Scheffé, H. 163, 245, 308
Sebastiani, P. 187, 302
Senn, S.J. 95, 308
Shah, K.R. 187, 308
Shao, J. 95, 301
Sholar, J.W. 303
Shrikhande, S.S. 70, 264, 300
Silvey, S.D. 301, 308
Simon, R. 122, 303
Singer, B.H. 37, 308
Sinha, B.K. 187, 308
Sloane, N.J.A. 162, 264, 304
Smith, C.A.B. 95, 304
Smith, H. 245, 302
Smith, K. 186, 308
Snell, E.J. 38, 83, 101, 122, 302
Spector, P. 269, 273, 297, 308

Speed, T.P. 245, 308
Stein, M. 190, 308
Steinberg, D.M. 163, 306
Stigler, S.M. 244, 308
Street, A.P. 15, 97, 264, 308
Street, D.J. 15, 97, 264, 308
Stufken, J. 162, 264, 304
Taguchi, G. 15
Taheri, J. 306
Tang, B. 126, 190, 301, 308
Taylor, D.W. 304
Teoh, S.K. 303
Thompson, M.E. 37, 308
Tibshirani, R.J. 15, 16, 307
Tsai, P.W. 163, 309
Tuck, M.G. 124, 125, 289, 309
Tukey, J.W. 247, 301, 309
Turnbull, B.W. 220, 304
Vaart, A.W. van der 245, 309
Vajda, S. 15, 309
Varney, A. 307
Venables, W. 270, 271, 290, 291, 296, 307
Verbyla, A.P. 220, 303
Verdinelli, I. 187, 300
Voss, D. 15, 62, 166, 302
Wald, A. 220, 309
Wang, J.C. 162, 309
Wang, W.K. 299
Wang, Y. 188, 190, 303
Wang, Y.-G. 220, 299
Ware, J. 224, 309
Wei, L.J. 223, 309
Weisberg, S. 245, 246, 301
Welch, B.L. 37, 309
Welch, W.J. 187, 299, 301, 308
Wermuth, N. 15, 302
Wesnes, K. 307
Wetherill, G.B. 222, 309
Whitaker, D. 304
Whitehead, J. 220, 309
Williams, E.J. 95, 309
Williams, E.R. 95, 97, 264, 304, 306

Williams, R.M. 220, 309
Wilson, E.B. 15, 309
Wilson, K.B. 14, 162, 300
Wolfowitz, J. 186, 305
Wong, W.-K. 187, 303, 304
Wu, C.F.J. 162, 309
Wynn, H.P. 163, 167, 168, 187, 305, 308, 310
Yates, F. 14, 36, 37, 62, 95, 96, 119, 122, 123, 162, 164, 165, 219, 264, 292, 310
Youden, W.J. 264, 310
Zhou, B. 97, 307
Zwischenberger, J.B. 299

Index

α designs, 81, 97
Analysis of covariance, 29–32, 57–61, 234–235
Ancillary statistic, 32
Autoregression, 207, 209–215, 220

Badgers, 215
Balance, 6–7
Balanced incomplete blocks, 70–80
    adjusted treatment means, 77, 280
    analysis, 76–80
        interblock, 79–80, 282
        intrablock, 76–80, 281
    and coding theory, 264
    construction, 72
    efficiency, 78, 214
    existence, 72
    and finite geometries, 260
    identities, 71
    incidence matrix, 71
    resolvable, 74
    in S-PLUS, 280–283
    sums of squares in, 77
    symmetric, 74
    and time trends, 213
Baseline variables, 4, 203
    adjusting for bias, 29–32
    and improved precision, 57–61
Best linear unbiased estimate, 227
Bias, see Systematic error
Binary response
    number of units, 199–200
Blocking
    methods of, 42
    and stratification, 161
Blocks, see Matched pairs, Randomized blocks, Latin squares, Balanced incomplete blocks
Brownian motion, 219

Carry-over effects, 88–94
    and Latin squares, 91–94
    two treatment-two period, 89–91
Case-control, 3, 16
Causality, 13–14, 34–36
Central composite design, 152
    rotatable, 152, 165
Combination of information, 11
Completely randomized design, 28–32
    randomization distribution, 38
    treatment allocation, 28
Components of variance, 196, 239–244
Computing packages, 267
Confounding
    double, 129–130, 164
    partial, 129
    simple, 127–129
    in split plots, 140–144
    in 3^5 factorial, 134
    in 2^k factorial, 127–130
Confounding variables, 34–35
Contrast subgroup, 114
Contrasts, 28, 53–57
    estimation, 53

mutually orthogonal, 54
    complete set, 55
  simple, 28
  in S-PLUS, 271
  sum of squares, 54
Cosets, 250
Cost of experimentation
  binary response, 199–200
  regression model, 197–199
  two treatments, 199
Counterfactual, 21, 34
Cross-classification and nesting, 144–149, 238–244

D–optimality, 174, 184
Degrees of freedom, 46, 102, 108, 133, 236, 240–242
Design measure, 170, 173
Design prior, 184
Design second moment, 170, 171, 173
Design sufficiency, 205
Dilution series, 178–180, 185–186
Double blind, 20

ED50, 209, 221
Effect modifiers, 107
Efficiency
  balanced incomplete blocks, 78
  of blocking, 52–53
  randomized blocks, 214
Enrichment entry, 204
Error correcting codes, 263–264
Error detecting codes, 263–264
Euler's conjecture, 257
Evolutionary operation, 163
Examples
  adaptive treatment allocation, 223
  balanced incomplete blocks, 83–85, 280–283
  cancer and nutrition, 120–122, 288–289
  central composite design, 166
  ECMO, 223
  electronics, 158–159, 163, 294–297
  factorial experiment, 101–102, 109–110, 285–288
    and blocking, 128–130
    three level, 134–136
    two level, 111–112
  flour milling, 124, 289–292
  fractional factorial, 118, 120–122, 131, 135, 142–143, 288–289
  growth of chick bones, 83–85, 280–283
  incomplete blocks, 85, 166, 283–285
  interblock analysis, 83
  pastry dough, 85, 166, 283–285
  potash, 50–51, 57, 273–279
  pulp preparation, 143–144, 292–294
  randomized block, 50–51, 101–102, 109–110, 273–279
  response surface methods, 289–292
  split plot, 142–144, 158–159, 292–294
  supersaturated, 139
  Taguchi methods, 158–159, 294–297
  weight of chicks, 101–102, 109–110, 285–288
Expected mean square, 45, 49, 146, 148, 235
Experiment, 1–14
  objectives of, 5–6
  scientific, 2
  series of, 11–12
  technological, 2
Experimental units, 4, 160
  choice of, 8
  distinct from observations, 8

INDEX   319

Factorial experiments
  advantages, 99
  complete, 99
  cross-classification and nesting,
    238–244
  prime number of levels,
    131–136
  in S-PLUS, 285–288
  two factors, 108–109, 144–149
  two level, see Two level
    factorials
Factors
  and category theory, 122
  homologous, 122
  nonspecific, 145
  qualitative, 100
  quantitative, 100, 107, 149–156
  random, 148–149
  setting levels, 100
Field
  definition, 254
  finite, see Galois field
Finite population sampling, 27
Fitted values, 228
Floating parameters, 237
Fractional factorials
  aliasing, 118–119
  in S-PLUS, 288–289
  three level, 134, 135
  two level, 116–122

G−optimality, 174
$GF(p)$, 132, 254
Galois field, 254–258
  construction, 254
  prime order, 254
General equivalence theorem, 174
General linear model, 226–244
Generalizability, 36
Generalized least squares, 246
Geometry
  finite Euclidean, 259
  finite projective, 258–260
    and incomplete blocks, 260

  linear model, 245
Graeco-Latin squares, 68–69
  10 × 10, 264
Gröbner basis, 168
Group, 249–253
  Abelian, 249
  definition, 249
  permutation, 251–253
    symmetric, 251
    transitive, 251
  prime power, 250–251
    generators, 250
Group divisible, 81

Hadamard matrix, 137, 261
Half normal plot, 115, 273, 288
Haphazard error
  reduction of, 41–43
  sources of, 41
Hat matrix, 238, 245

Illustrations
  agricultural field trial, 1, 8, 13,
    65, 87, 100, 146, 193, 202,
    215
  animal experiment, 58, 105
  causality, 13
  choice of treatment, 9
  choice of units, 4, 8, 9
  clinical trial, 1, 5, 20, 58, 88,
    94, 146, 202, 203
  community intervention, 87
  design objectives, 12
  ecological experiment, 215
  forming blocks, 43, 48
  genetic, 117
  Graeco-Latin square, 68
  intrinsic features, 100
  laboratory experiment, 100
  laboratory measurements, 20
  main effects and interaction,
    105
  matched pairs, 46

observational study, 2
opthalmology, 4, 46
precision improvement with baseline variables, 58
psychology, 65, 87
quality-quantity interaction, 153
randomization for concealment, 20
simulation, 181
social attitudes, 2
space-filling design, 181
spatial dependence, 215
stochastic modelling, 181
textile, 88, 210
time trend, 210
twins, 42
types of response, 5
Youden square, 75
Incomplete blocks, 70–85
    analysis, 82–83, 283–285
    and autoregression, 213
    connected, 82
    group divisible, 81
    partially balanced, 81
    resolvable, 81, 97
Intent to treat, 14
Interaction, 102–108, 112, 144–149
    components of, 131
    definitions of, 103
    higher order, 106
    interpretation, 106–108
    and main effects, 104–106
    qualitative, 105
    quality and quantity, 153, 166
    and random effects, 144–149
    sum of squares, 103
    and transformation, 107
Intrinsic features, 100

Latin hypercube, 134, 181, 187, 190
Latin squares, 65–70

analysis of covariance, 246
analysis of variance, 67
    balanced for carry-over effects, 91–94
    definition, 66
    linear model, 66
    orthogonal, 69, 132, 255–258
        example, 70
        existence, 257
        and lattice squares, 216
    orthogonal partition, 96, 258
    randomization, 66
    randomization analysis, 95
    in spatial design, 216–219
Lattice squares, 75, 216–217
Least squares
    in completely randomized design, 23, 30
    in general linear model, 227–232, 244
    geometry of, 230–235
    incomplete blocks, 76, 79
    in matched pairs, 44
Leverage, 246
Likelihood, 47, 64, 221, 226, 228
Linear sufficiency, 231
Log-linear model, 120, 288–289

Matched pairs, 43–47
    analysis of variance, 45–46
    binary response, 62
    least squares, 44
    linear model, 43–44
    modified, 46–47, 62
    randomization analysis, 44–45
Minimum aberration, 124
Minimum variance unbiased estimate, 227
Mixture designs, 154–156, 163

n of one designs, 95
Neighbour balance, 213, 214, 220
Noncompliance, 14, 35

INDEX 321

Normal probability plot, 115, 273, 288
Normal theory assumption, 23, 226
Number of units
  choice of, 193–201
    decision-theoretic, 197–201
    and power, 194–196
    and standard error, 194–196
    subsampling, 196–197
Number theory, 190

Observational studies, 1–4
  types, 3
One factor at a time, 124, 189
Optimal design, 169–191
  algorithms, 177–178
  Bayesian, 182–187
  check of linearity, 172
  dilution series, 178–180, 185, 187
  exact, 186–187
  generalized linear model, 180
  nonlinear, 178–180, 184
  quadratic regression, 176
  simple linear regression, 169–173
  types of, 177
Optimality
  types of, 188
Orthogonal arrays, 136–138, 158, 162
  asymmetric, 137
  construction, 162, 262
  definition, 136
  saturated, 137
Orthogonal polynomials, 56
  and time trends, 210
  in spatial designs, 218
Orthogonality, 6–7

Parametrization constraints, 105, 235–237
  in S-PLUS, 271
Partitioning sums of squares, 55–57
Permutation distribution, 27
Plaid design, 165
Play-the-winner rule, 205
Poisson process, 222
Principal block, 128
Proportion wrongly treated, 200

Quadratic residue, 255
Quasi-Monte Carlo, 188

Random effects, 148–149
Random walk, 213
Randomization, 19–20, 32–34
  and adaptive allocation, 206
  and ancillarity, 32
  biased coin, 206
  in clinical trials, 33–34
  of clusters, 39
  in Latin squares, 66
  and permutation group, 252–253
  restricted, 264
  uses of, 32–33
Randomization distribution
  and adaptive allocation, 206–208
  binary response, 38
  comparison of treatment and control, 37
  comparison of two treatments, 24–27
  completely randomized design, 38
  confidence limits, 26
  for efficiency calculation, 52
  Latin squares, 95
  matched pairs, 44–45
  randomized blocks, 50
  second moment properties, 25
Randomized blocks, 48–53

analysis of variance, 48–50
  efficiency, 52–53
  linear model, 48
  potash example, 50–51
  randomization analysis, 50
  in S-PLUS, 273–279
Rao-Hamming construction, 262
Recovery of interblock
  information, 47, 79–80
Reference mixture, 156
Residual effects, *see* carry-over
  effects
Residual maximum likelihood
  (REML), 243
Resolution, 118
Resolvable
  balanced incomplete block
    design, 74
  incomplete blocks, 81, 97
  Latin squares, 67
Response
  binary, 38, 62, 179, 199–200,
    205, 209
  choice of, 10
  definition of, 4
  intermediate, 5, 10
  surrogate, 38
Response surface methods,
  149–156, 162
  central composite design, 152
  centre points, 151
  design space, 151
  example, 289–292
  steepest ascent, 152
Retrospective adjustment for
  bias, 29–32
Robbins-Monro procedure, 209,
  221

Sample size, *see* Number of units
Second moment assumption, 23,
  226
Sequential experimentation,
  201–209, 219–220

Simplex centroid, 155
Simplex coordinates, 155
Space-filling designs, 172–173,
  181–182, 187, 190
Spatial coordinates, 217–219
Specificity, 36
Split plots, 140–144, 165
  examples, 141–144, 292–297
  subunit analysis, 141
  subunits, 140
  uses of, 142
  whole unit analysis, 141
  whole units, 140
Split unit, *see* Split plots
S-PLUS, 267–298
  analysis of variance, 270–271,
    276
  construction of designs, 273
  contrasts, 279
  data entry, 268
  data frame, 268
  fractional factorial, 289
  missing values in, 273
  model formulae, 270
  normal probability plot, 273
  parametrization constraints,
    276
  partitioning sums of squares,
    271, 278
  treatment means, 269, 275
Stagewise regression, 232–235
Staircase rule, 209
Stratification, 43, 160–162
Subgroup, 249
Sums of squares
  adjusted, 233
  contrast, 54
  general linear model, 228–232
  partitioning, 55–57, 109, 159
  in S-PLUS, 271, 278
  stagewise regression, 235
Supersaturated, 138–140, 162, 165
Synthesis of variance, 239

Taguchi methods, 157–159, 163, 167
  example, 294–297
Technical error, 21
Time series, 210–215
Transformation
  of quantitative factors, 149
  of response, 107
Treatment difference
  attenuation of, 203
  and baseline variables, 30–32
  linear model, 23
  modified matched pairs, 47
    likelihood analysis, 62
  randomization analysis, 24–27
    with carry-over effects, 90
  in time trend, 212
Treatments
  adaptive allocation, 204–208
  contrasts, see Contrasts
  definition of, 4
  quantitative factors, 56–57
  sequential allocation, 208–209
Tukey's test for non-additivity, 247
Two level factorials, 110–122, 127–130
  blocking, 127–130
  contrast group, 113–114
  estimation of contrasts, 115
  main effect in, 111
  treatment combinations, 111
  treatment group, 113–114
  Yates's algorithm, 123

Uniformly scattered, 190
Unit-treatment additivity, 21–22, 34, 39

Variance components, 47
Variance inflation, 60
Victorian England, 96
Virtue of abstinence, 140

Wreath product, 251, 253

Youden square, 74–75, 95